Design Participation

Sampsa Hyysalo

Design Participation

palgrave
macmillan

Sampsa Hyysalo
Design
Aalto University
Helsinki, Finland

ISBN 978-3-031-95436-8 ISBN 978-3-031-95437-5 (eBook)
https://doi.org/10.1007/978-3-031-95437-5

This work was supported by Aalto University.

Cover illustration: Cover artwork by Daniela Bianchi, on photos by Sampsa Hyysalo and Cindy Kohtala

This Palgrave Macmillan imprint is published by the registered company Springer Nature Switzerland AG
The registered company address is: Gewerbestrasse 11, 6330 Cham, Switzerland

If disposing of this product, please recycle the paper.

Virvelle ja Otavalle:
osallisilleni,
kokonaisilleni

Preface and Acknowledgments

This book is an offshoot of a long research program into how participation and multiparty collaboration is realized across different settings in design, peer activism, and innovation. When I started in the late 1990s, participatory design and user innovation were the games in town and still locked in holding workshops of various sorts. But soon there was a surge in open-source communities, codesign, living labs, end-user development, open design, company-hosted innovation communities, maker spaces, peer knowledge creation, citizen science, infrastructuring, and transition arenas (to name a few prominent ones). The shape of design participation has continued to evolve, and, importantly, it has been taken up by discipline and policy domain after another. What was once "new participation" has since become more mainstream as well as a dramatically more diverse envelope of participative engagements and forms.

The central tenet of this book is that participation matters particularly in environmental governance and transitions. But it only comes to matter if the matters of participation are duly attended to. To articulate both these aspects, the book draws from a wide portfolio of our realized design participation projects in environmental, cultural, healthcare, community, innovation, peer-to-peer, and transitions settings and is further nested in the social science studies of tens of others' exemplary projects that we have researched over the years. An interdisciplinary mix of design research, innovation studies, and science and technology studies (S&TS) has provided the orientation and theoretical means to study in detail what happens in these engagements, yet has been also importantly informed and qualified by our own design practice.

This basis has offered a unique opportunity to address the three key themes that intertwine across the book. The first theme is the import of the evolution of design participation: what it consists of and what the diversity of ways to pursue it currently offers for practice and research. The second theme is the neglected aspects of getting design participation done in practice, particularly drawing attention to the types of work that persistently surface regardless of how design participation is pursued. The final theme is how one can take one's bearings in this landscape through approximating the participation contexts and the suited ways to organize it therein.

The book integrates work that has been previously published in about 20 articles that have appeared in journals of different disciplines and have thus remained in regrettable disconnect prior to this volume. All these articles have been thoroughly reworked for the book to make the body text more accessible to practitioner and student audiences across design, planning, and public and environmental governance. At the same time, I have added layers of academic elaboration and bridging between the studies and vocabularies used in different disciplines for which there was no (literal or cognitive) space in the original articles. The many footnotes are there foremost for the benefit of academic and doctoral student readership.

Across the book, I mix "I" and "we." Most of the time "we" prevails as the "academic we" that lumps the author and audience into a rhetoric community. I use "I" for occasions where I want to be clear that an interpretation or position is likely mine alone and may not represent adequately the views of "actual we" that worked on the projects and earlier articles. It is this "actual we" I'm most grateful for making all the work and this book possible. My greatest thanks go to those original coauthors who traveled with me for several years in one or other sub-study track feeding into the book: Andrea Botero, Cindy Kohtala, Mikael Johnson, Tatu Marttila, Jouni Juntunen, Jack Whalen as well as my wife Virve Hyysalo have all been pivotal in developing the ideas, projects, reflections, and theoretical positions in this book. No less important have been the coauthors with whom I had the privilege to work on specific projects: Pia Hannukainen, Samuli Mäkinen, Louna Hakkarainen, Karoliina Auvinen, Sofi Perikangas, Sebastian Greger, Zagros Hatami, and Lotta Muurinen.

My special thanks go to all the over twenty INUSE research group members, present and past, for their great insights on the projects when they were in-the-making as well as on an early draft of its introduction. I've benefitted enormously from the careful commentaries to the next-to-ready manuscript chapters by Joanna Saad-Sulonen, Marika Silvikko de Villafranca, Andrea Botero, Cindy Kohtala, Torben Elgaard Jensen, and PJ Stappers. Daniela Bianchi did a fantastic job with all the images and figures, Mary Boss with language editing and Daniel Giacomelli with glossary and index. I wish to warmly thank Rachel Ballard and her team at Palgrave MacMillan for all the publisher side work.

Finally my very special thanks go all the participants in the design participation cases we outline in this book. The number of people who participated in person runs well over one thousand, and on top there are several thousand people who have anonymously participated digitally, so I have no means to extending a more personalized appreciation. The same goes for the approximately 600 students who have participated in my *values in design*, *strategic codesign*, and *design for social change* MA classes over the years—your interest and feedback has much shaped this book.

In terms of funding the Finnish Strategic Research Council grant Material Democracy—reducing polarization via material participation (Grant Number 365732) has made it possible to turn and extend the earlier works into a coherent book.

Regarding copyrights, whenever old text has been edited and not rewritten for this open-access book, it has either originally appeared under a creative commons CC-BY 4.0 license or I have author's subsequent publication rights to republish it. Even in the latter cases, I have avoided direct overlaps beyond 700 words to the original publication, as the pre-open-access era republication rights are currently a bit of a murky patch of copyright waters for open-access books. I have sought to reuse only such text passages originally wrote myself, save in Chapter 3 in which Andrea Botero and Cindy Kohtala appear as coauthors and have written parts of the text.

In itemized terms, the earlier versions of analyses that comprise Chapter 2 can be found in the following four articles: (1) Hyysalo, Sampsa, Kohtala, Cindy, Helminen, Pia, Mäkinen, Samuli, Miettinen, Virve and Muurinen, Lotta. (2014) Collaborative futuring with and by makers. *CoDesign* 10(3–4), 209–228; (2) Kohtala, Cindy & Hyysalo, Sampsa. (2015) Anticipated environmental sustainability of personal fabrication. *Journal of Cleaner Production* (99), 333–344; (3) Botero, Andrea, Hyysalo, Sampsa, Kohtala, Cindy and Whalen, Jack. (2020). Getting participatory design done: From methods and choices to translation work across constituent domains. *International Journal of Design* 14(2), 17–34; and (4) Hyysalo, Sampsa, Greger, Sebastian, & Hatami, Zagros. (2013) Co-designing the future health care. In (eds. Kivelä, Kaisa, Tyyri-Pohjonen, Sanna and Kinnunen, Krista,). *Living + For Better Living Environments.* Helsinki: Aalto ARTS Books.

Chapter 3 reworks two earlier articles and is coauthored with Cindy Kohtala and Andrea Botero, as the chapter uses our coauthored texts that they primarily wrote. Kohtala, Cindy, Hyysalo, Sampsa and Whalen, Jack. (2019) A taxonomy of users' active design engagement in twenty-first century. *Design Studies 67,* 27–54*;* Botero, Andrea and Hyysalo, Sampsa. (2013) Ageing together: Steps towards evolutionary codesign in everyday practices. *CoDesign: International Journal of CoCreation in Design and the Arts* 9(1), 37–54.

Chapter 4 builds on and extends two articles that we published under CC-BY-ND-NC to avoid our mid-range transition pathway toolset from becoming subject to commercial capture. Hyysalo, Sampsa, Marttila, Tatu, Perikangas, Sofi & Auvinen, Karoliina. (2019) Codesign for transitions governance: A mid-range pathway creation toolset for accelerating sociotechnical change. *Design Studies* 63, 181–203; Hyysalo, Sampsa, Marttila, Tatu, Perikangas, Sofi & Auvinen, Karoliina. (2019) Intermediate codesigning in transitions governance: Catalyzing and channeling participant action. *Design Journal.* All the coauthors have granted Palgrave (Springer) the right to use the text in this commercial open-access book.

Chapter 5 is a result of longstanding collaboration on Oodi Library with my wife Virve Hyysalo as well as with Louna Hakkarainen. Earlier versions on the analysis were published in Hyysalo, Virve & Hyysalo, Sampsa. (2018) Mundane and strategic work in collaborative design. *Design Issues* 34(3), 42–58; and Hyysalo, Sampsa, Hyysalo, Virve and Hakkarainen, Louna. (2019). The work of democratized design in setting-up a hosted citizen-designer community. *International Journal of Design* 13(1), 69–82. The latter part of the chapter is a rewriting of an article we wrote with Mikael Johnson and Samuli Mäkinen. Mäkinen, Samuli, Hyysalo, Sampsa & Johnson, Mikael. (2018) Ecologies of user knowledge: linking user insight in organizations to specific projects. *Technology Analysis & Strategic Management* 31(3), 340–355.

Chapter 6 continues my longstanding collaboration with Mikael Johnson, published originally under CC-BY 4.0, albeit extended and edited for the present book. Hyysalo, Sampsa & Johnson, Mikael. (2024) Making sense of methods and approaches to user involvement. *The Design Journal*, 1–29.

Competing Interests The author has no competing interests to declare that are relevant to the content of this manuscript.

Contents

About the Author

Sampsa Hyysalo (Ph.D., Dos) is a professor of Codesign at the Department of Design, Aalto University, Finland and professor II of Science & Technology Studies at NTNU, Norway. His interdisciplinary profile bridges between design, science & technology studies, innovation studies, and sustainability transitions. He has investigated for over two decades how design and participation can be arranged alternatively and to what effects. This has meant engagement in for instances, participatory design, codesign, user innovation, living labs, open design, maker spaces, peer knowledge creation, user communities, design communities, and citizen science. These engagements have been pursued both in empirical studies as well as through own interventionist design research. They have resulted in not just new methodologies but in a deep understanding on how different designer-user arrangements play out in sociotechnical change. Domain-wise much of his recent work has focused on citizen involvement with sustainability transitions and design participation in the public sector. Sampsa has published over 100 peer reviewed articles and book chapters. He has previously authored seven books, the latest being *Citizen Activities in Energy Transition: User Innovation, new communities, and the Shaping of Sustainable Future* (Routledge, 2021), *The New Production of Users: Changing Innovation Communities and Involvement Strategies* (with Elgaard Jensen and Oudshoorn, Routledge, 2016). He was awarded the Academy of Finland price for social impact of research in 2010, EASST Freeman Price 2016, and Aalto university's excellence and social impact prices in 2017 and a researcher award in 2024. He has given a range of keynote addresses, including 4S and EASST conference presidential plenary talks, and is part of Design Research Society's International Advisory Council.

Co-contributors to Chapter 3

Andrea Botero is a professor of new frontiers in design in department of design, Aalto University, Finland. Her work engages with the possibilities and contradictions of participating in design, and designing participation today. She has explored these issues in multiple domains ranging from everyday life to scientific endeavors. Through her research, she aim to understand how collectives (broadly speaking) come to understand the design spaces available to them and how designers could infrastructure those spaces better, especially looking to support livable futures. Her work draws from and happens at the intersection of participatory design, science and technology studies, and transformative creative practice.

Cindy Kohtala is professor of design for sustainability in Umeå institute of Design, Umeå University, Sweden. Her research bridges Design Research and Science & Technology Studies (STS). She is interested in how active citizens experiment with sustainability hands-on and materially, whether through hacking and making in fab labs and makerspaces, citizen science, urban activist projects, or other grassroots initiatives aimed at local knowledge sharing and empowerment. She studies industrial transition at the local scale: the stories we create about localizing design, technology, and production and the emerging practices around peer-to-peer participation. She examines how activists build infrastructures for alternative ways to produce and consume and how they involve wider publics in imagining and enacting alternative sustainability futures. In UmArts she explores ways researchers, designers, crafters, and activists can collaborate on knowledge-making in sustainability transformation using design-led methods.

Acronyms and Abbreviations

#lm	#lovemilla, which is the name of an online drama series
API	Application Programming Interface
CeLib	Central Library
CRM	Customer Relationship Management
DIY	Do-It-Yourself
ELY	Regional Business and Environment Authorities
EU	European Union
f2f	Face-to-face
FCL	Friends of Central Library
FLOSS	Free/Libre/Open Source Software
FNBC	Finnish National Broadcasting Company
HCD	Human-Centered Design
HCI	Human-Computer Interaction
IPR	Intellectual Property Rights
ISD	Information Systems Development
IT	Information Technology
LUW	Lead-User Workshop
MLP	Multi level perspective
MLS	Minimum Lovable Service
MTPT	Mid-Range Transition Pathways Toolset
MVP	Minimum Viable Product
MVS	Minimum Viable Service
NGOs	Non-Governmental Organizations
OSS	Open-Source Development
PD	Participatory Design
PR	Public Relations
PV	Photovoltaics
Q&A	Question and Answer
RET/RETs	Renewable Energy Technologies
S&TS	Science and Technology Studies
S-RET	Small-scale Renewable Energy Technologies
TA	Transition Arena

TM	Transition management
UX	User Experience
WDC	World Design Capital

List of Figures

List of Tables

CHAPTER 1

Design Participation for Social and Environmental Change

Introduction: From Design Solutions to Design Participation

Our era is one facing increasing resource depletion, climate change, growing overpopulation, biodiversity loss, and many other grave side effects of our present human condition. Social and environmental change is today something designers should engage in if they are to walk the talk on the transformative potential of design. "Changing existing conditions to preferred ones," in Simon's (1969) famous adage, entails acknowledging that designed objects, services, and procedures do not exist in an environmental or social vacuum. Designs are shaped by a range of contextual conditions and, by the same token, hold potential to change those conditions.

Pursuing design in a thought-through manner is a quest that may well define what design is and what its relevance may become in the current century (Friedman et al. 2002; Löfgren and Stolterman 2002). Yet, traditional form-giving and understanding of immediate human–material and human–human interactions may fall short of the adequate analysis and courses of action required to bring change to complex social and material circumstances. But what is the alternative? There is a line of design manifestos proposing demanding reorientations for designers as a condition for a shift to something more important than rounding the corners in service of narrow commercial and governmental interest (e.g., Papanek 1972; Fry 2008; Irwin 2015). Yet these envisioned overhauls may set unrealistic demands on designers, expecting them to now become able environmental scientists, systems analysts, political economists, psychologists, and sociologists. We may thus need to wait a while to see if these demands bestowed on designers become realizable and result in new more adequately oriented design disciplines.

S. Hyysalo, *Design Participation*,
https://doi.org/10.1007/978-3-031-95437-5_1

Meanwhile there already are concrete, doable, and demonstrated ways by which design can become a major contributor to social and environmental change. *These ways rest in shifting design from a creative act of defining and solving problems to developing ways by which others can do so.* Together with these others—such as activists, entrepreneurs, civil servants, and politicians, as well as sociologists, political economists, psychologists, and systems analysts—fewer capacities and academic decathlete skillsets are needed from any one actor. The requisite knowledge and resources to act on social and environmental issues in particular domains can be collectively found, even if the challenges would be daunting for any individual or professional group. This is the path of design participation toward environmental and social change that the present book outlines. And it is just as relevant for designers seeking to work participatively as it is for other professionals and activists seeking to employ participative approaches from design to address their concerns. Let us concretize this with Vignette 1:

Vignette 1: *Let us consider the import of Victor Papanek's "Design for the Real World" (1972) for climate change mitigation. Papanek famously made the case for how designers must shift from being part of the problem in environmental sustainability to becoming part of the solution. At its early time of writing, there were still a limited amount of exemplary sustainable design solutions, and among the better ones of his examples is the "artificial burrs" project by J. Herold and J. Truan to revegetate arid areas. It was an early regenerative biomimicry design that used bio-degradable plastic shaped into burrs (plant structures that have hooks in the edges to snare in animals or in other vegetation if carried around by wind). The burrs were coated with seeds and growth boosting solution so the seeds can start germinating in nutrient-rich material once the rains arrive, in turn making roots and binding soil and sand in place, helping other vegetation take hold. The artificial burrs were designed to be deployed from a plane in thousands and present the kind of design solution that would presently be needed* en masse *to mitigate biodiversity loss and climate change. Yet we may ask, will any amount of such designs be enough to change the unsustainable structure of production and consumption that produces, for instance, many aridity problems, loss of vegetation, and its capacity to reduce carbon dioxide from the atmosphere? Countermeasure design solutions alone are unlikely to change the root causes that keep on damaging the environment.*

What are needed are complementary efforts to transform wide and often deeply rooted sociotechnical change. These typically go beyond what any individual designer can hope to achieve by designing material or service solutions. An example of a design participation solution that has targeted such wider change with some measure of success is the creation of a mid-range transition pathway toolkit (MTPT) (Hyysalo, Marttila et al. 2019) that brings a diverse group of change makers into a transition arena (Loorbach and Rotmans 2010) to work out how systemic change in a sector such as energy can be realized in a given city, area, or national context. This toolkit has helped participants to envision pathways that have contributed, for instance, to end coal use in energy production in

Finland by 2025 in contrast to the "earliest feasible 2035," which prevailed prior to the first energy transition arena we ran in 2017.

The idea of steering transitions via *activating stakeholders originated from Dutch innovation scholars, but new design tools and procedures were needed to make this way of collective envisioning and changemaking "travel" successfully in a new country context and to duly address the acute changes needed in the short and mid-range timespans (Heiskanen et al. 2009; Lähteenoja et al. 2023). We discuss the design of these tools later in this introduction and in Chapter* 4, *but for now the takeaway is this: designing for other actors to take action is a crucially important complement to what designers can ever hope to achieve on their own.*

Because both the environment and democracy are currently under attack in many places, it is necessary to remember that environmentally relevant activities as well as collaborative design practices are not something that are hypothetical or external to us. We are always already entangled in design processes with environmental consequences and implications for democracy. It is necessary to be aware of the considerable technical and cultural apparatuses that have already been created to facilitate ever-more efficient and comprehensive participatory processes. Because of this, we need to understand the felicitous circumstances under which design participation can both be a positive "democratic" experience and be mobilized to have positive environmental impacts. The mission of the book is to explore and shine light on this path, which is already there, but which is now more necessary than ever. Thus, better understanding of design participation—in principle and in practice—is important to today's designers and planners, and it is just as relevant for the other professionals, disciplines, and stakeholders who engage people in collective efforts toward social and environmental change. This is to say, design participation can be made to matter, and to this end, the matters of design participation need elaboration.

From Participatory Design to Design Participation

Collaborative approaches to design, planning, innovation, and social change are no longer fringe activities. They form a diverse and broad envelope that includes formalized public engagement initiatives, participatory design (PD) projects of all kinds, industry user involvement, and peer-to-peer communities in a variety of citizen domains (Simonsen and Robertson 2013; Ehn et al. 2014; Botero and Saad Sulonen 2023; von Hippel 2005, 2016; Bogers et al. 2010). This means that user participation in design, together with collaborative design endeavors by active users, have become part and parcel of how contemporary design and innovation operate (Hyysalo and Jensen et al. 2016; von Hippel 2016).

In a previous book, "The New Production of Users" (Hyysalo and Jensen et al. 2016), we elaborated how users' participation in design and innovation processes is no longer such a maverick pursuit with inherently empowering underpinnings, as it may have been in the previous millennium. Participation by citizens and customers has also become an object of industrial strategizing like any other organizational activity. Some of the new champions of the techniques, methods, and organizing of design collaboration are, for instance, large software firms such as SAP and Oracle that are single-mindedly profit-oriented (e.g., Pollock et al. 2016; Mozaffar 2016). Yet these companies do care about democratic opportunities created for their customers and their potential empowerment in design and innovation. They must care about it to entice users to participate, which, in turn, offers these companies important resources through identifying development opportunities, gaining free development efforts and marketing, and staying in tune with how the customer industries are developing (Pollock and Hyysalo 2014; Mozaffar 2016; Pollock et al. 2016). At the same time, these corporate giants involve their users only insofar as it creates value for the company and fosters loyalty among the communities of their customers. This is the further mainstreamed state of the "new landscape" of codesign (Sanders and Stappers 2008). It could be seen to result from "the creativity of designers and people not trained in design working together in the design development process," as Sanders and Stappers (2008) define, and outline how it offers potent new ways to organize design activities. The related matters of participation are far from simple, though.

The traditional response to mainstreaming participation by users and stakeholders in design and innovation has been to view mainstreamed codesign and co-creation activities as poor cousins of proper PD (e.g., Robertson and Simonssen 2012; Palmås and Busch 2023). A relatively loose definition of codesign, such as that offered by Sanders and Stappers (2008), could in this view mean any involvement of non-design-trained people, any creativity, and just informing designers. At the extreme, this could mean sucking out every last drop of value from the customers for corporate benefit (Thrift 2006; Hyysalo, Jensen et al. 2016; Palmås and Busch 2023). This is to say, while there is nothing wrong with codesign being good business as well, its procedures and methods *in themselves* ensure only a narrowly democratized design (Suchman 1995; Hyysalo, Jensen et al. 2016).

PD emerged in the 1970s to enhance workplace democracy and "is about the direct involvement of people in the codesign of technologies they use. Its primary concern is how collaborative design processes can be driven by the participation of the people who will be affected by the technology that is being designed," as Simonsen and Robertson (2013, p. xix) put it. PD seeks to involve the "problem owners" as well as those people that would become implicated, indirectly affected one way or another, by the design endeavor. It "actively engages people, groups and other actors in collaborative processes to explore and co-create their everyday technologies, practices and environments," as Smith et al. (2025, p. 1) condense it. It seeks to create adequate conditions for direct participation in design and decision-making for these people, not limiting these aspects of design for professional designers. It has excelled at creating means and procedures to make meaningful participation in design possible (Asaro, 2000; Voss et al. 2009; Simonsen and Robertson 2013). Contemporary PD has placed increasing emphasis on human and planetary needs, global diversity, and social equity (Smith et al. 2025).

Just as importantly, for designers, PD entails accessing and participating in the real problems that are faced, and fundamentally defined by, the problem owners (Bjerknes et al. 1987; Greenbaum and Kyng 1991; Voss et al. 2009; Simonsen and Robertson 2013). Most PD practitioners have been vehement about progressive aims and siding with underprivileged parties as preconditions for PD, as well as the need to connect PD endeavors to wider sociotechnical change (Bjerkness et al. 1987; Bødger et al. 2017). The "P" in PD also stands for political (Beck 2002; Smith et al. 2025), and there is no escape from political choices being present beyond the commitment to participation.

These PD premises are a healthy antidote to the instrumentalization and exploitation of people in a clever guise of participation (Thrift 2006; Palmås and Busch 2023). After all, participation holds an aura of innate goodness in democratic societies and thus provides a strong lure for people to join, whatever the cause they joined factually involves (V. Hyysalo 2022).

Yet, the wide uptake of collaborative and participative ways in design and delivery of services has raised new challenges for PD and codesign, regardless of one's orientation (Hyysalo, Jensen et al. 2016; Smith et al. 2025). PD projects, principles, and ways of organizing have become a subset of deep and well-justified design participation, with different casts of roles and responsibilities than what PDP has honed for designers, users, and other stakeholders (Hyysalo and Johnson, 2024; see Chapter 6).

First, open-source software and hardware and content communities such as Wikimedia, Linux, and MySQL have arguably taken democratically governed design further than any designer–user collaboration ever before (e.g., Ratto 2004; Ardati, 2023). After all, any professional designer involved in those collectives is just one participant among many. Open-source development communities hold a wide range of organizing and governance principles, only some of which adhere to the democratic ideals of PD. The same goes for different hacking and hacker collectives, citizen science initiatives, and peer-to-peer design networks. Also, design communities and living labs "hosted" by companies and the public sector can pursue considerable depth and autonomy in design participation and shared service delivery with well-justified yet less ambitious and strict criteria for participation (e.g., Bovaird 2007; Ansell and Torfing 2014; Hyysalo et al. 2018). Contemporary PD further underscores the importance of global diversity of communities and cultures in design endeavors, including a wide variety of communal design engagements (Usenyuk et al. 2022; Huybrechts et al. 2025).

Second, democratic participation opportunities are just one of the substantive values of importance in almost any design project that seeks to make a difference in the real world. Raising it above other substantive values becomes increasingly problematic when design participation is pursued for substantive projects and changes (Friedman et al. 2002; Borning et al. 2005; Friedman et al. 2013; Friedman and Hendry, 2019). This becomes accentuated when the prime concern is not with human participants who can stand for themselves and their peers (Poikolainen et al. 2025). For instance, if the aim of design participation is to create arrangements by which to mitigate advancing biodiversity loss, biodiversity preservation clearly is at least as important substantive value as democratic participation opportunities created for the participants. Principles for participation may, in fact, occasionally become mere "means" or be just one aspect of the "mix of ends" being pursued (Friedman et al. 2002; Mok and Hyysalo 2018). Phrased more provocatively: over-privileging participation over other substance values runs a risk of enrolling people into "participating in participation" as a surrogate for affecting the issues and matters involved (Kelty 2020).

Third, once participation moves from the predominantly relatively small-scale PD projects to large-scale projects that are to touch the lives of millions of people, the diversity in both preferred and attainable participations expands (e.g., Dalsgaard et al. 2012; Dalsgaard and Eriksson 2013; Hyysalo and Hyysalo 2018). Confining participation only to a select few, who can participate intensively and invest time to have a direct say over design decisions, is not evidently the best way to deal with the wide diversity of preferences, voices, and contexts that affect such design endeavors. There is a real danger of lapsing into over-privileging those social groups, which designers are implicitly most concerned about, if the less interested, more temporally fleeting, content-wise shallow, and less determining forms of participation are cut out because they are not qualified as "proper enough" participation (cf. Arnstein 1969; IAP 2023).

Vignette 2: *Let us consider how design participation may (have to) change in large-scale projects through the case of the Helsinki flagship downtown library, Oodi. It featured large-scale participatory efforts to develop the uses, spaces, and services of the €100 million undertaking that sought to define what libraries of the next era would be. Twelve different kinds of collaborative design activities were carried out during its two-year concept design phase. These included in-depth participation forms such as establishing a designer–user community, a pilot maker facility, a lead-user workshop, and participatory budgeting. Many of these forms were subsequently taken up at a wider scale in the City of Helsinki. But just as importantly, the project collected ideas and "library dreams" from the public* via *open online and offline channels, sent its staff to meet the public in urban events to gain further ideas, reached out to citizens with bicycle-based mini-libraries, transformed all of Helsinki's 30 branch libraries into design spaces for a few days, and ran a workshop series with partners, stakeholder groups, and some user groups (Hyysalo and Hyysalo 2018; Hyysalo, Hyysalo et al. 2019; V. Hyysalo 2022). These more fleeting and superficial engagements could easily be* *discarded as limited to low-rung participation, at best merely "informing" planners (Anrstein 1969), and one could question if they really adhere to any PD principles.*

Yet these more superficial engagements allowed well over 3,000 people to express their ideas, perspectives, and feedback on the project in contrast to the few hundreds that could participate in more depth (itself an unusually large number of participants in a PD project). Most importantly, the overall "participation mix" that was built up could reach out to a great diversity of people, most of whom would not have been able or willing to participate in any one form of participation, particularly to the more in-depth and time-consuming ones (Hyysalo and Hyysalo 2018; V. Hyysalo 2022; see Chapter 5). Moreover, academic projects aside, there are typically multiple competing and well-justified "native" understandings of what democratizing design should mean. In the Oodi participation activities, empowering citizens and underrepresented customer groups, shifting design work and decisions to users, and better informing decision-making by the elected council of the City of Helsinki were all rationales that could not be jettisoned or privileged over one another if the participation activities were to have legitimacy and impact on the planning (Hyysalo, Hyysalo et al. 2019) (Fig. 1.1).

All this calls for navigating between the perhaps too loose and instrumental way to think about design participation in codesign (cf. Sanders and Stappers 2008) and perhaps too restrictive way to think about it in PD (cf. Robertson and Simonsen 2012; Smith et al. 2025). Both stand as important beacons for navigating the waters of the broader phenomenon of *design participation, the active engagement of affected and interested parties in shaping design, and the direction of socio-techno-environmental change*. Before we can give design participation a more precise definition, we need to endeavor deeper into its substance to get a better sense of what it means to pursue it "for the real

Fig. 1.1 Oodi Central Library, Helsinki, Finland (Photo by Kuvio, reprinted with the City of Helsinki permission)

world" (to paraphrase Papanek 1971): for real problems, in earnest, and in a skilled manner.

From Professional Designers to Those Who Engage People in Design Participation

Public participation is pursued across administrative and political domains (Bovaird 2008; Bovaird and Loeffler 2012; Ansell & Torfing, 2014). It is commonly carried out by civil servants, consultants, and researchers that have their backgrounds in social and political sciences or in the substance fields where the participation takes place, such as in healthcare, transport engineering, or cultural fields (Bherer et al., 2017). These professionals typically have competency in running the legally mandated and organizationally routine forms of participation in the area they work in, such as public consultation, hearing procedures, customer surveys, focus groups, or a Delphi-method study (cf. Kelty 2020; Bherer et al., 2017). In recent years, many such people have realized that these established tools for the job—most dating from the 1930s to 1970s (see Chapter 6)—provide somewhat limited forms to engage with the interested publics. Many have further realized that the need for novel ways to engage people has become heightened in processes that materially transform our societies: design participation is de facto dealt with by a wider range of professions than just designers and planners (cf. Marres 2012; Hyysalo, Jensen et al. 2016; Chilvers and Kearnes 2016; Kelty 2020).[1]

Furthermore, this widened range of professionals are increasingly faced with citizen activists and collectives who contest and complement expert plans with viable and well-elaborated alternatives such as open design initiatives and urban shadow plans. Activists often get amazing things done and muster impressive social organizing, but just as often their achievements come with

[1] Kelty (2020) argues that since the 1960s there has been a radical expansion in participation across the public sector, giving rise to domain-specific participation routines as well as cohorts of professionals whose prime expertise rests in organizing and running participation across different fields rather than augmenting the interplay between affected citizens and public administration. In a cynical view, this development has rendered ***public participation*** into a legislated and technical set of procedures and arrangements in routine governance. A more optimistic view is that while such cynicism is warranted—much public participation veers toward limited influence—the participation routines within organizations as well as the cohorts of participation experts precisely generate interest and legitimacy for efforts to pursue more encompassing and ambitious forms of participation. The interactive expertise generated (Collins and Evans 2002) is a necessary springboard for envisioning and practically realizing better-suited forms of participation. Moreover, actors such as public-sector consultants and service design agencies are ***intermediary actors*** who engage in participation in enough settings in a sustained length of time that allows skill and insight to cumulate, be refined, and be disseminated between different domains and projects (Howells 2006; Stewart and Hyysalo 2008; Kivimaa et al. 2018; Hyysalo et al. 2022). They are highly important actors, many of whom go out of their way to foster and improve the participatory practices they are involved in even though there are tensions involved too.

high time and resource expenses and use of whatever means they happen to be acquainted with. Thus, a better understanding of how design participation works and what it may offer and produce (and how it may not work and what it is unlikely to provide) would often be helpful. It is also important to recognize that democracy activists—not just academics, companies, or public administrations—develop many tools, repositories, and platforms that aid design participation. Strengthening their appraisal of design participation may further nurture this two-way street.

Design participation is equally relevant for managers and executives, who make strategic decisions about what kinds of customers and public relations their companies or public organizations pursue. As pointed out above, world-leading companies are strategically managing not only their customer relations and brand value but also the creative potential of their current and potential customers in innovating, designing, testing, marketing, tutoring, troubleshooting, and repairing their offerings (Hyysalo, Jensen et al. 2016; Pollock et al. 2016; Prandelli et al. 2008; Sawhney et al. 2005; Antorini et al. 2012). Understanding how design participation works helps to nurture and structure these contributions to lasting mutual benefit of the company and its clientele.

Thus, while this book often uses the word "designers" for those who initiate, plan, coordinate, run, and act on the results of design participation, it should rather (by now) be taken as a proxy for all those people who end up engaging other people through design participation, be they activists, entrepreneurs, politicians, planners, managers, or civil servants in some public-sector domain. And getting these diverse audiences onto the same page entails the next insight: all need a better understanding of how design participation is *enacted in practice*. This leads us to design participation *work*, which remains ill conveyed amidst reporting new methods and principles for participation (Hyysalo et al. 2019; Botero et al. 2020), and this is where we turn next.

From Participatory Ideals to Design Participation Work

The ideals and methods of collaborative design tend to steal attention from carefully produced knowledge of what it takes to realize participation beyond the ideals that have motivated it. "For real" in this book directs attention to the "pragmatics" of design participation in messy real-world settings—often of pivotal importance to whatever is eventually realized of the participations (Jensen and Petersen 2016; Botero et al. 2020). While there are some examples of rich guidance of how particular participatory methods play out in practice, these foremost accompany single methods or techniques in a how-to orientation (e.g., Sleswijk-Visser et al. 2005; Sanders and Stappers 2012) rather than reflect on the import of pragmatics to design participatory endeavors and principles.

Vignette 3: *Let us consider the importance of reporting the mundane pragmatics through an interesting and exemplary study case of pluriversal design to restore Kamaru Stream in New Zealand in conjunction with granting it with the same legal rights as human beings. Barcham (2022) engaged in "pluriversal ontological design in practice," drawing from Escobar (2018). The focus of his article centers on how multiple ontological imaginaries can be brought into negotiation and design. After the initial phase of cultural mapping and fishing surveys, he writes how "[w]e used a relatively standard codesign process of facilitated meetings around specific goals to formulate the actual design of the management plan—as required by the management plan's template and as specified in the guiding legislation. The drafting of the various aspects of each required section occurred over the six-week cycle. Pre-readings were sent out to working committee members to help them understand the goals of the upcoming working meeting. Then, when we convened as a group during the 10 months we worked together, we used a broad range of codesign techniques. The important point to note is not the techniques,* per se, *but how they were deployed. In undertaking each codesign session, we prioritized Ngāti Hori [clan living in the area] knowledge and ways of being in the use of each codesign tool, technique, or methodology. For example, whenever we worked in smaller groups or pairs, the different ontologies always were represented in each group. Thus, all participants were always engaging with someone from Ngāti Hori in their work. The process of designing, building out, and using the various boundary objects promoted a form of infrastructuring through which a new nascent imaginary began to emerge. The worlds of the various individuals from the different participating groups—Māori, non-Māori, and Ngāti Hori—began to come together. This imaginary was a decolonial imaginary, in that it was centered on the role that Ngāti Hori played as kaitieki (guardians) of the river and the ontology upon which this role was based, even as it engaged with ideas and concepts from Western science and related ontologies" (Barcham 2022, p. 13).*

Barcham (2022) underscores that what mattered was *how* codesign techniques were deployed, not the techniques per se, and particularly the practice of always representing different ontologies of each group. This is a core lesson to remember. Still, because of the word-length limit and reviewer request as to how to detail the process in a manner that is short but palatable to journals' audiences,[2] this one paragraph is what we learn about how the "relatively standard codesign process of facilitated meetings" was operationalized through "a broad range of codesign techniques" in which their novel deployment was really what mattered, and only some aspects of which are then detailed later in the article. For the reader interested in how participation actually unfolded in-practice, this one-paragraph description of a 10-month process with several six-week cycles provides a trailer rather than the movie of all that must have been involved in this decolonial deployment of codesign. The reader is left craving for all that was done when the codesign techniques

[2] Personal communication, Manuhuia Barcham, 28 February 2025.

were operationalized, adjusted, altered, and facilitated to work in this radically different setting. What might be the constituents of the "relatively standard codesign process" and how its form, make-up, phases, facilitation procedures, and timeframe (and so on) affected the process and its outcomes? How did the broad range of techniques beyond the mapping tools and decolonial procedures outlined complement each other, or were there also tensions between them that required tailoring? If so, how and where? What competencies were required of the facilitators and from the participants to enter and proceed with this exemplary pluriversal design? How did all these (and many other) aspects interrelate in making the codesign succeed?

Now we could have used almost any project description from the history of participatory and collaborative design instead of the Kamaru Stream example above. As in the Kamaru Stream article, where the details of how codesign techniques were operationalized were forsaken for article length and reviewers' view of what was most important in the article, the actual practice of design participation remains a blind spot across the available handbooks, research literature, and theory base that is used to inform design participation (for the same argument, see Jensen and Petersen 2016; Wilkie 2010; Hyysalo and Hyysalo 2018; Botero et al. 2020).[3]

When the project pragmatics and the interrelations between design participation principles, methods, and operationalizations are lacking, students and practitioners tend to find it difficult to think through what different forms of participation or even methods might be suited and what they entail in practice. There are now hundreds of different techniques and methods that have been used in codesign and PD projects (Muller and Kuhn 1993; Muller and Druin 2012) and considerably more in neighboring fields that could be adjusted for these ends (e.g., participedia.net; peoplepowered.org; www.actioncatalogue.eu). As Barcham (2022) wisely underscores, it is not the method or technique per se, but how it is deployed, adjusted, paired, and linked that makes design participation prosper. There is an emerging cohort of studies at the intersection of design and science and technology studies (S&TS) that insists on conveying this concreteness on which we build on in this book.[4] Next we elaborate how, similar to pluriversal design, the different deployment

[3] The likely reasons behind such vagueness are manifold. Beyond the publication word limits and reviewers not recognizing the importance of detailing the pragmatics, it is demanding and often hectic to pursue design participation and the meticulous recording that would then be needed to reconstruct how the processes unfolded tend not to be done. It would often require added academic resources and interest to do that, as all sociologists and anthropologists of real-life technology projects well know (e.g., Sørensen and Williams 2002; Bobrow and Whalen 2011; Szymanski and Whalen 2011). Furthermore, careful analysis of such data beyond post-project reflections by the practitioners is very time consuming and thus tends not to be done (Verheig et al. 2016; Jensen and Petersen 2016).

[4] Some of the important studies here include Jensen and Petersen (2016); Wilkie (2010); Botero et al. (2020), Woolrych et al. (2011), Johnson et al. (2014), Verheig et al. (2016), Jensen (2012).

can transform the rationales and principles by which design participation is organized and practically realized.

From Concept Design Workshops to Active Use and Design-in-Use

The last two decades have witnessed a fundamental expansion in the understanding of what it means to participate in design and innovation. Most design engagement is not organized by designers during concept design but carried out by the people in their everyday home and work lives. Without the investment of time and interest and resources people devote into making designs work for them, designs soon become obsolete dead chunks of matter or dead bytes in the hard drives.

Let us be a little more specific about this. Perhaps because designers were so long engrained in working within industry conditions, the study and critique of how designs were used fell onto psychologists, sociologists, and ethnographers studying workplaces and homes. And what they gradually discovered was that, in those settings, "a design" as it was "designed" may be an exception rather than the rule (Silverstone et al. 1992; Poole and de Sanctis 1994; Miller and Slater 2007; McLaughlin et al. 2002). Collectors aside, nobody wants a design. What people want are tools, services, and objects that allow them to get things done by suiting their everyday life or work practices, their settings, and their identities.[5] Of course some designs match the needs and preferences of some users "as designed," and the fit of many has been greatly improved by human-centered design and UX (ever since Gould and Lewis 1985; Norman and Draper 1986). The great diversity among adopters, adoption contexts, and adoption moments, however, means that most designs do not adequately outright match the needs of many users. People are able to use the designs nonetheless but only after domesticating them. This process is then a precondition for the actual usability, utility, and appreciation of goods and services-in-practice (Berker et al. 2005; McLaughlin et al. 2002; Alter 2006; Hartmann 2023).

The ensuing process can be comprehensive as in information systems adoptions, in which over 30 different "adoption moves" have been observed (Poole and DeSanctis 1994). It can result in extremely diminished "versions" of designs (Mol 2002; Hyysalo 2007), as in the case of nurses using less than 7 of the 52 alarm and messaging functions of a medical device (Hyysalo,

[5] Surely, "a design" is an important moment in the biography of an artifact (Kopytoff 1986; Hyysalo 2010), but it covers strictly speaking three moments. The first is when an artifact reaches the end of the production line and becomes neatly packaged. The second occurs when an artifact becomes exhibited in a museum or in home decoration. The third occurs in between when a discrete piece of design is used exactly for its intended usage(s). But this "use as is" has turned out to be less the default case than designers (and design theorists and historians) have assumed (e.g., Crilly et al. 2008), as we discuss in Chapter 3.

2007). Minimally, designs are appropriated into the moral economy of a household or a workplace, incorporated into activities carried out and into functioning assemblies of other artifacts, objectified within the meaning and significance structures of these environments as well as converted to convey social meanings and status of their owner rather than just that of their designer (Silverstone et al. 1992; Berker et al. 2005; Paavilainen et al. 2017). All in all, the "adoption" of an artifact is characterized by a shift from one economic, social, technical, and cultural setting to another and in so doing altering the relevant characteristics, uses, meanings, and surroundings of "the design" (Engeström and Escalante 1995; Hasu 2001; Hyysalo 2010; Szymanski and Whalen 2011).

Recognizing that much design-related activity happens during use and that this provides possibilities to shift into design-in-use participation arrangements has marked an expansion in the landscape of design participation that has remained hitherto only partially recognized. Codesign and PD have traditionally focused on concept design phases, which indeed are the prime locus of collaboration if one is working under the conditions of mass manufacturing, as it becomes expensive to modify the design after launch (Sanders and Stappers 2008).[6] Concept design and "fuzzy front end of innovation" have been emphasized also to have real influence in industrial organizations: historically, the knowledge and concerns about users moved slowly from after-sales and complaints to marketing to usability testing to concept testing to idea generation and to strategy.[7]

But the wave of natively digital goods and digital-physical manufacturing systems have allowed shifting of design into use time, as the production can proceed through evolutionary releases based on adjustable product/service architecture (Baldwin and von Hippel 2011; Vezzoli, Kohtala et al. 2018; Abel et al. 2011; Kohtala et al. 2020). When this is the case, the upfront design can be only a starting point ("minimum x" as in minimum viable product, minimum lovable product, minimum viable system, lean start-up, and so on)

[6] As early as in the 1970s, there was a proliferation of numerically and then digitally controlled flexible production that could produce several variations of the good and be altered for design iterations, similar to IT systems of the era. Consequently, the pre-design and design-in-use begin to find an increasing amount of interactivity such that designers at the producer organization would want to actively gauge and solicit the designs by users for incorporation in future product generations (Gardiner and Rothwell 1985; Fleck 1993; Lundvall 1985). The cost-efficiencies of production still pushed flexible manufacturing to strive as much as possible toward "right design and designing it right" from the onset. This is the lingering manufacturing backdrop in much industrial design theory and teaching still today, placing the industrial designer as the creative interpretative point of passage between the producer and the customer. Once mass-customization and digitally controlled production lines proliferated in earnest after the turn of the millennium, this cast of roles began to unravel as designers were now involved in designing configuration systems that enable the customers to make their own designs from a selection of alternatives (Johnson et al. 2014; West and Piller 2014).

[7] Thanks to PJ van Stappers for this observation and formulation.

from which evolutionary designing begins for finding the solutions and forms that adopters value.

This change allows design participation to expand in time from pre-design to design across continuous releases, to expand in space from a design of a single service or product to (additional) designs of parallel configurations at different customer sites. It further allows design participation to expand in diversity as adopters can do both low-level and extensive designing to many aspects of the system to suit their needs and preferences. At the extreme, the expansion of design participation can result in users taking collective control of the design process altogether as we know from open source and open design movements (Abel et al. 2011; Botero 2013; Kohtala et al. 2020; Botero and Saad-Sulonen 2022; von Hippel 2016).

Vignette 4: *Let us consider how design participation has changed from mass manufacturing to design-in-use architectures by revisiting the "classic" article "Codesign and the New Landscape of Cocreation" by Sanders and Stappers (2008). Writing in the late 2000s, they outlined how codesign and cocreation have become the new black in design and innovation and how design professionals would now need to collaborate, liaise, and recognize the design contributions by, in their words, non-designers. They elaborate on the process of codesign by departing from Damien Newton's squiggle diagram about the reduction of uncertainty in design to depict moves between professional design and users in an iterative manner with lessening amplitude from establishing design criteria to product launch, at which point the collaboration seizes (Fig. 1.2, upper part).*

But soon after this assessment, the "new landscape" shifted yet again. The early design-in-use approaches such as co-realization, meta-design, and end-user development had showed that it is difficult not just for designers but also for users to elaborate what all the new systems and services should do, as the new systems open further possibilities for action, and that design-in-use strategies could cater to the virtuous design evolution that could produce high-quality designs (Hartswood et al. 2002; Fischer 2003; Lieberman et al. 2006). In the 2010s, these early trials gave way to wide industry uptake in the wake of proliferation of digital services and agile development.

What this meant was that once a flexible service or system architecture was in place, the initial release of the service or product no longer needed to be closed for production and to present a break in design participation. The initial release could be designed for open exploration and evolution in mind, include design seeds that users might grow, engage users to reconfigure and make design iterations, and involve an extended process and channels through which users could elaborate back to the designers about how the system could be further improved. With this radically different baseline idea about design participation and who is to drive the design process and when, the cycles of design–launch–use and the roles reserved for professional designers and other actors could be shifted, as users could take responsibilities in design along the development of their own work and procedures (expressed in squiggles in the lower part of Fig. 1.2 and in a fuller version in Chapter 3 with a connection to infrastructuring).

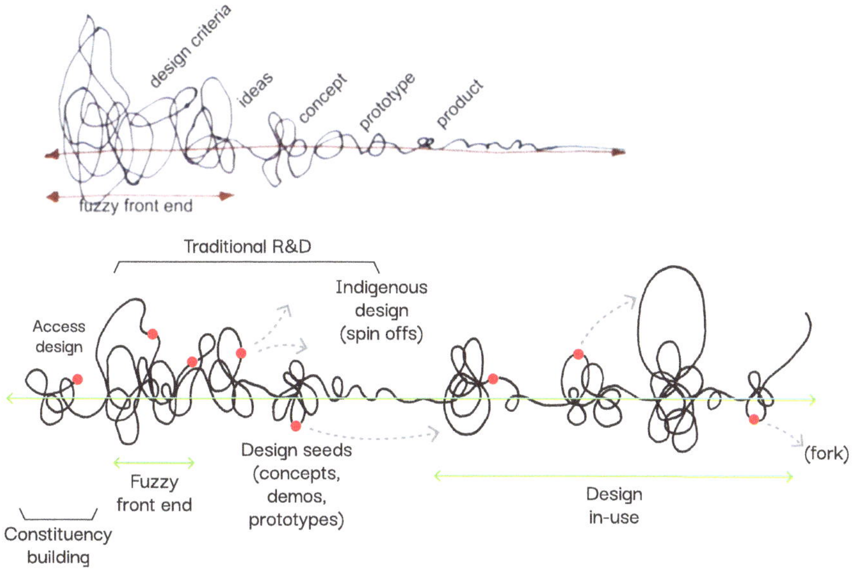

Fig. 1.2 Design participation premised on product launch and design-in-use expressed in design squiggle remakes by Sanders and Stappers (2008) and Botero and Hyysalo (2013), drawing attention to different temporal organization and orientation to uncertainty. The upper image Sanders and Stappers 2008, reprinted with permission

To summarize the design-in-use and manufacturing system insight: the golden age of designer-centered mass and flexible manufacturing was just one era and one contextual condition for design. Presently, many societal and industrial domains feature far more distributed design agency as their baseline condition because of its clear benefits. This opens up possibilities for organizing design participation differently in time and space, as we discuss in Chapters 3 and 6.

From Invited Participation to Invested Participation

From the previous, it follows that sociotechnical[8] change seldom ensues from "a design" that "diffuses" to change society (Rogers 2010). Instead, a much broader, longer, and windier process ensues. Shifts in the material make-up of designs and related users' and other actors' practices happen almost always when an innovation moves from early trials to further sites (see, e.g., Hyysalo, Pollock et al. 2019; Glaser et al. 2021; Wigboldus et al. 2016; Sengers et al. 2021; Ryghaug and Skjøsvold 2021). New uses, new adopter groups, and contexts of use emerge as well, alongside effects by and changes to regulations that tend to accompany often decades-long processes of sociotechnical change (Flichy 2007a, b; Hyysalo 2010, 2021). In all, sociotechnical change is a long social process that necessarily reaches beyond any initial "social design," as in designing primarily to address social issues (Kimbell and Julier 2014) or any initial technical design solution, however conductive of wider changes it then may become.

The net effect is what Ryghaug and Skjøsvold (2021, pp. 65–66) call "distributed participation" in sociotechnical change that recognizes that designs, be they technological or social initiatives, are necessarily spread more widely in time and space than just their initial development contexts (Kohtala et al. 2020; Hyysalo and Juntunen 2024). A great many users and other affected and implicated actors come to participate, and whether they play along, resist, opt out, alter, or voice their views in public or to developers then defines their participation. As part of this, design and technology development can create new kinds of publics that can investigate and test not just the technology projects but society as well (Marres 2012; Ryghaug and Skjøsvold 2021).

Be the participants partaking in old or new communities, their capacity to participate hinges on this wider distributed engagement with the matters, communities, and issues (Miettinen and Hasu 2002; Hyysalo 2003; 2010; Marres 2012; Kelty 2020). This accentuates the need to move beyond thinking participation in terms of something that designers, planners, or other organizers "invite" (Callon 1999; Irvin et al. 2013; Ryghaug and Skjølsvold 2021, pp. 64–65). While invited participation offers a way to delineate the participants along various desired characteristics such as demographic and local representativeness and can be used to detach the participants from the group and self-interests and achieve "contributory autonomy" (Kelty 2020), such

[8] Sociotechnical change is a way to say that, on a closer view, next to all "social" phenomena are materially constructed, and equally, "technical" matters cannot be separated from social action taken in and around design and in and around use and disposal. Insisting on a sociotechnical view to change is fitting for design as a vehicle and orientation to change, as design is, perhaps rare exceptions aside, predicated on changing both humans and materialities (vs. an opinion campaign or demonstration march for change). The use of sociotechnical change instead of technical and social is also a way to anchor the discussion into particular theory traditions that underline how change always proceeds from existing (sociotechnical) conditions to new one(s) (not ex-nihilo) and never in a sociotechnical vacuum (Williams and Edge 1996; Clarke and Star 2003; Geels 2002).

detachment is limited by three dynamics that are critically accentuated in *design* participation.

The first of these dynamics was introduced above in noting how design participation is critical for embedding the change initiative in the site and *domain-specific details and context relations* that can be hard for designers to approximate and anticipate. The participants thus bring into the proverbial table an understanding embedded in their community, workplace, or everyday life. Such understanding is tied to collective matters and easily gets lost or remains lacking in invitational participant selection premised on single autonomous individuals from different social settings (Greenbaum and Kyng 1991; Kelty 2020).

The second dynamic lies in the participant's ability to elaborate the embedded knowledge and particularly the ability to anticipate what might be possible, available, and desirable directions of change as discussed in the previous section. This is somehow not an innate characteristic of humans. It is related to specific *practices* and domains and a deepening understanding of *relations* between many people, things, and infrastructures—to put it bluntly, in design participation, it is not so much the attributes of individual participants that matter per se but the participant as a "carrier of practice" capable of articulating matters in their work or everyday community (Hyysalo 2003, 2010; Shove et al. 2012; Alter 2006).

The third, and related, dynamic is that participation in design hinges on the ability to elaborate concerns and preferences *relevant to design* and not only generally. While this can be greatly supported by tools, templates, information packages, and facilitation, the elaborated-ness of design aspects tends to be something that needs to be rehearsed and experimented on to become refined, often through first-hand material engagements. This is to say, there is a considerable difference between general elaborated-ness of collective concerns that a participant is able to voice and the elaborated-ness of design issues related to those concerns (Miettinen and Hasu 2002; von Hippel 1994). The effects of this embedment in community and material engagement have been found to be profound. They have, however, gone partly unrecognized because they are best elaborated not in participation or design-related literatures but in innovation studies (von Hippel 1976, 2005, 2016). The baseline finding is that there is an elaboration and transfer barrier between design and use domains to the effect that in many settings it is users, not producers, who end up innovating and who hold the best awareness of the available design avenues and their implications (von Hippel 2005, 2016). Furthermore, the combination of use domain and design domain knowledge and expertise requires considerable effort and time to build, regardless of if this is by accumulation of experiences in two domains or through purposeful efforts—and tends to thus be highly concentrated to very few people in any pairing of use and design domains (Franke et al. 2006; von Hippel & von Krogh, 2016).

For traditional forms of invited participation—be they focus groups, codesign workshops, townhall meetings, mini-publics, or whatever—the implication is that participant delineation techniques that work well for slow changing, well-rehearsed, and generic matters tend to produce superficial and poorly informed views and limited design import in more uncertain emerging and technology-intensive matters.[9] At the same time, the advanced elaboration and design capacities that many *invested people* hold offer possibilities to move from just knowledge elicitation (Lezaun & Sønerud 2007) to direct *coproduction* of design outputs and decision-making (Bovaird & Elken, 2012; Miller & Wyborn, 2020), as participants can directly design and be involved in in-depth considerations in expertise-requiring decision-making. Particularly when their competencies and efforts can be pooled, participating people can often surpass the technical, planning, and design competencies by specialized experts (Baldwin and von Hippel 2012; Prandelli et al. 2008; Bovaird and Loeffer, 2012). Due attention must thus be paid to a "bottom-up" emergence of design participation in a given domain as a precursor to, a complement to, and a contestant to forms of participation organized by firms or public administrations.

Vignette 5: *Let us consider the bottom-up design participation and community forms that tend to remain invisible to designers and other public or private company actors through user innovation in heat pumps. The ongoing transformation toward low-carbon energy systems has resulted in rapid adoption of heat pumps for keeping homes warm. A heat pump is basically a refrigerator operating in reverse. One might thus assume that people install and run heat pumps just as much "as is" as they do their fridges. Yet once we investigated the use of heat pumps among Finnish citizens, it turned out that tens of thousands of people had implemented add-ons or simple hacks to their devices, such as aesthetically more pleasing housings for the outdoor units or simple plastic receptacles to gather condensed moisture so it can be easily skipped even after it has frozen solid in the Finnish winter. More encompassing hacks and repurposing was done in hundreds of projects, which featured 113 new-to-the-world innovations that had either a new function or material realization (Hyysalo et al. 2017; Hyysalo 2021). And the peer-to-peer design participation is not limited to direct hacking and redesign. In the Finnish heat pump case, citizens have set up internet communities that have featured not only hundreds of thousands of posts and 200 million reads but also very in-depth, thought out, timely, and well-curated advice for the selection, purchase, running, monitoring, and improving*

[9] Another way of phrasing this is that there is principled uncertainty about (socio)material transformation: its objects and their effects are not stable and fully preconceived entities but relational ones, realized only in their entanglements and thus with varying effects, both potentials and harms, in different settings and times (Engeström and Escalante 1996; Orlikowski 2002; Hyysalo 2004). Just as well, the actors and collectives implicated and brought into being by sociotechnical change can be difficult to delineate beforehand exhaustively (Latour 2005; Marres 2012), even as some clearly affected actors evidently can be, see Chapters 3 and 7.

of renewables in different settings. All this has made a crucial contribution to the development and proliferation of this technology that initially featured all kinds of glitches and uncertainties in its new cold-climate context (Hyysalo et al. 2018; Hyysalo 2021; Hyysalo and Juntunen 2024).

At the same time, prior to the publication of our findings, the heat pump resellers and public actors in the energy sector were oblivious of, or neglected, this grassroots phenomenon and its importance. None of this should be surprising. In representative surveys on modifications and user innovations in households, 1.5–6.1% of consumers report having innovated in their homes in the past three years, and in many countries such as the UK, the time and expenditures devoted to user innovations far outweighed innovations by the country's private companies. And it has been regularly found that innovations and communities by citizens are systematically belittled as if they were "hidden in plain sight," owing much to how citizens do not have a need to market their actions in public or toward the experts, but only use it to their own benefit (von Hippel 2016).

The dependency on investment in the domain and practice specifics in design participation give it a particularly uneasy footing regarding "the problem of participation," as Kelty (2020) phrases it. The experience of participation always features, at once, hopes of empowerment and fears of cooptation. If design participation must feature somewhat (or even highly) invested participation, what guarantees may there be that it is not highjacked to only furthering the narrow self-interest of design-able groups or participants that are in no way elected or legitimately representative of all the impacted people? As importantly, if more ambitious forms of design participation must proceed via people known to be (somewhat) invested, is there not always a legitimate remit for its organizers to act as the final decision-makers in the name of balanced public (or corporate) good, in turn, rendering all but fully independent voluntary design communities necessarily devoid of final decision-making power?

These are deep-seated quandaries. Yet we need to make a recourse back to why design participation is not only about its principles—it is not something that can be settled at a desk or in an armchair. As Jensen phrases it (2016), the involvement of people in design tends to "straddle" any one normative position or concern and form an "ambiguous middle ground," where both its organizers and participants have to take their footing, for example, between invested and disinterested participation; between empowerment and cooptation; between community embedment and representativeness; between broad domain knowing and design ability; between top-down and bottom-up modes of organizing and control; between general positions and their (muddled) detailed realization; between organic and expert participation; and between well-elaborated issues and uncertain objectual relations (Hyysalo, Jensen et al. 2016,b).

Thus, in this cautious but more optimistic register, due acknowledgment of distributed participation shifts the focus of how participation is invited, coordinated, and organized by developers to a more general issue of how the participation evolves in ecologies of other actors and how the change is then enacted across the diversity of actors involved (Ryghaug and Skjøsvold, 2021 p. 66; Hyysalo et al. 2022; Hyysalo and Juntunen 2024). For design participation in our present era, a key big question remains of how design can address and move from the traditional invited participation and from a project-focused and project-confined model to supporting and laying grounds for the different forms that participation can take. Extended design-in-use, formation of peer-design communities, design for social innovation, and infrastructuring (Benkler 2006; Björvigson et al. 2012; Simonsen et al. 2020) are the kinds of efforts that bridge between these and arguably present an important current in the present landscape of co-creation, to rephrase Sanders and Stappers's (2008) reasoning from two decades ago. And nowhere is this shift more acutely needed than in efforts to deal with wide structurally embedded environmental problems in consumption and production.

From Designable Change to Catalyzing Actions into Motion to Achieve Sociotechnical Change

All too often new sustainable design concepts, circular product-service systems, and initiatives for behavioral change have become incremental or trampled over by the inert logic of consumption and production in the sectors they seek to transform. It has become evident that in many key societal sectors, single alternative design solutions simply cannot compete against the inertia created by path dependencies that have resulted in "sociotechnical regimes" being built over decades by interlinkages in industry structures and production technologies, investment patterns, scientific bases, institutions and policies, market mechanisms, and cultures of consumption (Köhler et al. 2019).

Against this backdrop, society-wide long-term changes are not "designable" per se as they result from intertwined actions that span regulation, technology development, altered consumer practices, taxation, and new business creation (and so on). They require different types of actions by different types of actors. While the unsustainable patterns of production and consumption within these systems cannot be remedied by single designs, there is a heightened need to better connect the relevant actors needed for bringing about such wide-cast societal changes in liberal democracies: decision-makers, experts, civil servants, citizens, NGOs, and business leaders (to name but a few). Thus, the envisioning and coordinating hundreds of complementary actions presents a new avenue for sustainable design.

The idea of contributing to long-term societal transitions is promising for designers engaged in sustainable design who seek more encompassing intervention strategies beyond industrial and service designs that appear ineffective and too limited in both scope and temporal orientation. Consequently, several authors have put forward program proposals such as designing for transitions (Ceschin and Gaziulusoy 2019) and transition design (Irwin et al. 2015). As this kind of project requires substantial investment of time and resources and new ways of design engagement, there are still only few realized projects and tested approaches globally (Ceschin and Gaziulusoy 2019). One of these is the development of a mid-range transition pathway toolset and arenas which, by now, have been deployed in 16 different contexts, each lasting several months.

Vignette 6: *Let us consider what it might take to realize the efforts to design for transitional change through the example of the mid-range transition pathway toolkit (MTPT). The MTPT was built from cross-breeding transition management (Loorbach and Rothmans 2006) and material and social support for direct participation in PD (Bødger et al. 2004). Its aim is to clarify the mid-range transition's vision and goals and to produce concrete mid-range pathways, as well as to build capacity for groups of 15–50 change makers and problem owners with diverse and complementary competencies and perspectives that allow them to develop viable alternatives together within an area or sector. The groups go through a facilitated process to determine change goals and the sociotechnical pathways that will lead to them from the present state. To aid doing so, the MTPT has a set of predefined forms and categories that participants use to add content and arrange the actions to be taken either in face-to-face settings or digital co-presence (*see Fig. 1.3*). The work typically proceeds both from the present day forward and from the time of the mid-range goal backward. The results are typically comprised of a few key clusters of pathway steps, usually totaling 15–25 steps and 40–70 identified actions that are needed to support those steps, with identification of knowledge and action gaps in the road to the envisioned mid-range goal. The use of PD techniques allows the condensing of the process to 4–7 half days with in-between digital commentary and follow-up. Such versatility is important as time use is the key to attracting busy people such as leading politicians, activists, and business leaders. The mid-range time span of 10–15 years and the ensuing higher concreteness in actions have been important developments, as widely agreed visions of a sustainable future are no longer in short supply like they were when transition management emerged in the last millennium. For instance, when the Finnish Prime Minister's office commissioned us to run an MTPT arena for UN Agenda 2023 implementation in Finland, the UN sustainable development goals formed the point of reference shared by the participants (Lähteenoja et al. 2021). Often experiments toward change are typically afoot, and anchoring MTPT arenas to ongoing local processes has been observed to foster higher legitimacy. At best, MTPT processes provide new ideas of how changes can be pursued and offer a counterbalance against incumbent actors who tend to dominate policymaking arenas and political lobbying.*

Wide sociotechnical change calls for new sensitivities in managing wider and more diverse participant groups. This includes new criteria and means for enlisting the participants, arrangements for bridging knowledge gaps between participants who peer learn from each other (rather than learning primarily between participants and designers or organizers), and working with new types of focal issues (such as regulatory changes and subsidy schemes) and interest groups than has been customary. These consequently call for devising adequate "intermediate designs"—the means, tools, and procedures that help

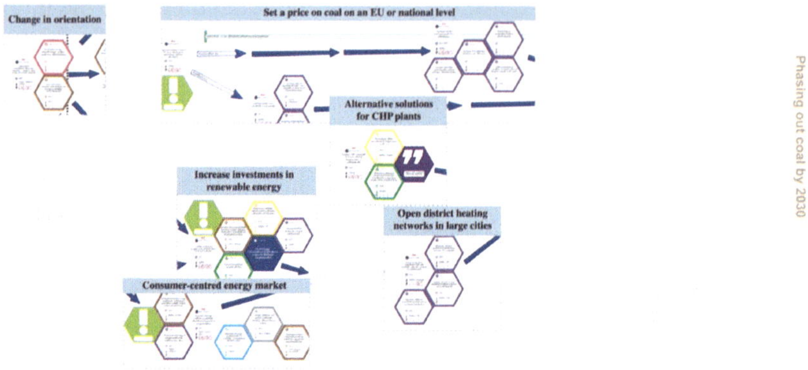

Fig. 1.3 An example of a completed pathway, with five key clusters of action systemically leading to ending the energy use of black coal in Finland by 2030

participants reach meaningful outcomes—suited for these new contexts. How to arrive at designs that are apt for catalyzing participant actions without affecting the participant's say over the outcomes with different expert and citizen groups is not a trivial issue (Hyysalo et al. 2019b; Lähteenoja et al. 2023).

Another way to phrase this is that the expansion in the scope of design participation has also meant that it has expanded beyond designable aims—public deliberation, knowledge coproduction, and participatory culture creation—forming a sub-part of larger planning or reform processes (Bovaird and Loeffler 2012). In such endeavors, there is a limit as to what can, and even should, be designed and what is to be designed *for* creating arrangements and spaces in the hopes that impacted people take ownership of them or otherwise make good use of designer efforts (Botero 2013; Hyysalo and Hyysalo 2018; Hyysalo, Marttila et al. 2019; Hyysalo, Perikangas et al. 2019). In other words, this is an example of professional design having shifted to becoming a catalyst of distributed participation in social change.

Defining Design Participation

With these insights, we are now in the position to begin to define our take on design participation. Participation has long been a central topic in political science, taken to underlie democratic governance of societies. Participation is thus typically addressed from the point of view of what the attributes of good participation are or what they should be. For instance, representativeness, inclusion, ensured quality of deliberation, participant learning, detachment from personal and group interests, informed opinion formation by autonomous individuals, and contributory autonomy are typically taken to be fixed virtues that universally apply to all settings and times of participation (cf. Renn et al. 1995; Dryzek 2000). Chilvers and Kearns (2019) argue that holding up these virtues of proper participation, de facto, entail that participation must take place in discrete settings in which the organizers can ensure the virtues are duly realized, and the (assumedly pre-existing) publics thus get to engage with the issue at hand through prescribed formats, techniques, and procedures that ensure high-quality participation.[10] The question of how such participation then produces design outputs and wider societal effects is rendered secondary to the legitimacy of the participation organized.

Closer to design participation are positions that focus on some aspect of participation, such as the depth of control and decision-making power that participants hold, for example, in "participation ladders" (Arnstein 1969; IAP, 2023), taxonomies of citizen involvement in public service design and delivery (e.g., Bovaird 2007; Bovaird and Loeffer 2012), and theories of who should or could effectively engage in design and innovation and on what grounds and with what expected outcomes (von Hippel 2005, 2016; Schuler and Namioka 1993; Simonsen and Robertson 2013; Smith et al. 2025).This group of theories does not, however, cohere into a unified body, and most of them focus on one aspect or one kind of design participation. Consequently, the implications of these different substance theories about participation are often in direct conflict with each other (as we discuss in Chapters 2, 3, and 6). As Kelty (2020) poignantly demonstrates, participation presents an enduring problem across disciplines and across decades that evades attempts to contain it into one form or another.

While appreciative of these resources to think about participation,[11] the work we directly build on in this book has departed from examining design and technology-related participation-in-practice. It has underscored how the registers of participation in political science are foremost prescriptive idealizations that become coproduced in participatory arrangements with a range of just as consequential material and social elements (Jensen and Petersen 2016;

[10] Form (Hyysalo et al. 2016a) and (procedural) Format (Chilvers & Longhurst, 2016) are from here on used partially interchangeably. In our use "form" includes a somewhat wider configuration of setting in which participation takes place than just the procedural format. In our use "format" conveys the repeated staging of participation in only slightly varying ways akin to TV formats (cf. Clausen et al. 2020), thus being a somewhat wider, materialized and often less formalized counterpart to "participatory methods." If used as a verb "formatting" participation it is used it in a sense of arranging participation into a format, not with a connotation to formatting computer hard-drive, that is, rendering the sociotechnical setting(s) in which participation can happen as amenable to some but not other formats.

[11] As the father of action research Kurt Lewin remarked "there is nothing as practical as a good theory." Theories sensitize to issues that ought to be considered, provide conceptual understanding to make sense of the complex issues involved, help establish basic parameters that the solutions must meet, and aid making sense of the often-messy results (and much more). What Lewin implies is that if there is an applicable, tried, and tested theory, it can significantly inform in preparing for the action, conducting it, interpreting it, and learning from it for the next instances.

Hyysalo, Jensen et al. 2016; Chilvers and Kearnes 2016; Hyysalo and Hyysalo 2018). In this view, normative characterizations of participation are important cognitive and legitimating *resources* to the realization of design participation (Hyysalo, Hyysalo et al. 2019). Yet participation, as a consequential practice, is not reducible to these idealizations or their specific proceduralizations (Voß and Amelung 2016; Kelty 2020). Drawing from constructivist S&TS tradition, Chilvers and Kearnes (2018) characterize participation as an always *emergent* coproduced outcome of subjects (organizers, participants, publics), objects (issues, social ends), and formats (procedures, facilitation) in heterogeneous collective practices through which publics engage in collective issues (Chilvers et al. 2018, 2022). Similarly, Hyysalo et al. (2016) underscore the *produced* nature of productive participants through the build-up of new appealing *forms* for participation and through these forms the ever-new *matters* to which diverse people can engage with and (co)produce outcomes as well as new participating subjectivities, in turn legitimating the efforts to build and run ever-new forms by which people can participate.

As we indicated in the second section of this chapter, *design participation* means *acts that shape or transform design, planning, or innovation in sociotechnical change pursued by the potentially affected actors.* This is foremost a restrictive definition to designate specific kinds of participation issues that importantly differ from other kinds of participation and regarding other kinds of matters. This definition entails that the outputs of the design participation are designs or plans, alterations to designs, or interim steps taken toward designs, plans, or progressing sociotechnical developments that are to be realized (whether they then actually materialize owes to a further set of influences).

Design participation thus also differs from generic public participation in running or monitoring institutions and citizen or civil society participation in service provision, which is common across public services from hospitals and schools to NGOs and neighborhood associations (Bovaird 2007).

As such, design participation tangents what Marres (2013) describes as material participation,[12] but it would include routine material participation only insofar as it means taking active part in socio-material change (Ryghaug

[12] Marres (2013) elaborates two senses of material participation. Material participation points to a general dimension often forgotten but important in all participation, and it may be transformative to how informal, linguistic, and discursive dimensions of participation are conceived (stressed by a line of works by Latour and Callon). Her own further take is to elaborate on material participation as a specific form of participation that happens through engaging with materialities in a way that forms or contributes to publics, which is particularly evident in relation to issues that affect not just public life but everyday private action by citizens. The publics brought about through material engagements are at once a poor fit to the Aristotelian ideal of proper public participation, requiring disentanglement from one's own private interests and everyday concerns. They "should not bear the marks of the influence of the setting" (p. 28) and have their own dynamics and problems as publics, as they can be argued to be ephemeral and fluid in comparison to ideology or interest-based publics, owing to the multivalence of material objects and their changing relational composition. In my reading some ambiguity remains in her account as to how

et al. 2018; Marres 2012; Hyysalo 2021). To concretize, somebody shifting from a petroleum car to an electric vehicle (EV) or installing a heat pump can be taken to respond to climate change through material participation, particularly if displaying these consumption choices in public (cf. Marres 2012). To also become design participation, one would need to alter the equipment or daily practices around them somehow or file improvement suggestions to manufacturers, or participate in peer-to-peer exchanges about how to buy, install, fix, or optimize them to the effect that they facilitate changes in their own or other's installations. Thus, material participation in the mode of "mere" consumption, however made public, such as driving an EV without meddling with the car or its uses or its meanings in public discourses, would remain a liminal case of design participation. At the same time, non-public acts of design and making would count as design participation as long as it had sociotechnical effects, such as a heat pump working better and longer and resulting in lower energy consumption and a longer replacement cycle (cf. Hyysalo 2021; Kohtala et al. 2020).

Also design participation is not deliberative participation to gather or formulate only public *opinion* on some matter or issue, even technological issues, if it stops at opinion formulation (Renn et al. 1995; cf. Chilvers et al. 2018). If cut off from the prospect of resulting in any (further) transformative output in the matters deliberated on, it remains participation in public opinion formation rather than design participation.

Our take on design participation resonates with those in contemporary PD who are reluctant to follow traditional PD in closing what constitutes "good" participation in different sectors and global contexts, emphasizing how participations produce designs in different practices and spaces (Smith et al. 2025; Huybrechts et al. 2025). Yet our take on design participation also includes the more instrumental participation practices in design pursued in companies and public organizations as well as in normative pursuits for other social ends such as environmental improvement.[13]

explicitly or implicitly public this specific material participation ought to be to count as participation.

[13] Observing that "non-human" actors and "more-than-human design" have entered the recent parlance and efforts to redefine PD (Smith et al. 2025; Poikolainen et al. 2025), a clarification of the position of this book in this regard is in order. The relational ontologies developed in S&TS are at the back of the way I address design participation. They are premised on agency being an emergent property of an assemblage of elements, which the Cartesian or Kantian ontology would dichotomize into human, social, biological, physical, and so on (Latour 1987, 2005). An early polemic expression of this stance was Latour's proposal to give due agency to "non-human" actors in the composition of "social" and "human," to rattle the inherent Cartesianism and residue modernism in social theory. The recent design critiques evoking non-humans, and the need to move beyond human-centered design through "more-than-human (centered) design" is in this light highly welcome in bringing to the front the need to (de)center design efforts to the needs of other species, habitats, and related materialities. As with Latour's non-humans, the more-than-human design works well as a polemic. Yet it is not a coincidence that less polemic but more carefully elaborated concepts have largely replaced non-humans in S&TS.

Thus, for actions to count as design participation, they must concretize into contributions toward altering or conceptualizing designs or (other) forms of socio-material organization that affect change. Such engagements hold their very own quirks and turns that may be only rather weakly present in some general public participation settings. As argued through this introductory chapter, this transformative substance begs for specific and close-up attention, which an overly general theory and definition will not duly attend.

These themes of design participation will be deepened and widened in the chapters as follows.

Chapter 2: Design Participation Work and the Workshop as Its Microcosm.

Design participation in today's world requires more than implementing methods, following general principles or operationalizing normative stances. Besides skills in selecting and implementing codesign methods, there is a larger repertoire of issues that need attention, if one takes the promises and limits of participation seriously. To gain a better understanding of what is involved, we introduce six types of work that go into participatory and community forms of design: framing work, relevance work, selection work, constituency building work, intermediate design work, collaboration work, and outcome work. The chapter introduces these work types through the most common and at least seemingly most straightforward arrangement in designing participation: the co-located workshop. To this end, the chapter analyzes two one-day workshops that feature different aims, contexts, and underpinnings. The first of the workshops is an envisioning workshop for future maker spaces for the flagship Oodi Central Library, which eventually became a lead-user workshop because of the demanding knowledge requisites, long-time horizon, and solution and trend knowledge that was to be coproduced in the workshop. The second workshop is a concept design workshop for a design project to build an information portal for Finnish maternity wards. This workshop is in many respects a more classic participatory workshop situated in the middle phases of a design project. It enables multiple interacting user groups to ideate, define, and make decisions about a design concept, mediated by a wireframe prototype, facilitator, and results from an earlier field study. These two workshops present two ends to design participation: the latter one that is relatively straightforward to plan and run and an effective way to proceed in the design project, and the other that is complex, demanding, and resource-intensive to plan and carry

As Latour's (1993) messenger lectures already eloquently outline, there is no such thing as a human existing detached from a wide variety of our material and biological constituents. Historically, it makes sense to talk about humans only much after these entanglements have become constitutive to our evolution, some 1–3 million years ago, making human existence relational and agency always a relational outcome of an assemblage of actors involved in an event (Latour 2005). Hence, regarding terminology my preference is for relational over non-human. This is particularly so if the "centered" is dropped from more-than-human-centered-design, the remaining "more-than-human" at once re-essentializes "human" to something it never was, as well as begs to consider some if not all entities that are not part of that (re)essentialized human. This then can basically mean anything and everything, making it a practical impossibility to "center.".

out yet produces a process and outcomes that would be very difficult to get to by any other means.

Chapter 3: Extending Participation into Use Time And Designing-in-Use (coauthored with Andrea Botero and Cindy Kohtala).

Chapter 3 expands our journey beyond the enactment of design participation. The last three decades have witnessed suites of design approaches, both in research and practice, that are premised on further design being carried out by people using the products and services. The advantages of continuing collaborative design after implementation were first proposed in the 1990s, and this insight was later turned into design-in-use approaches such as co-realization, meta-design, and aging together. The rise of hacking and making has, in turn, led to open design initiatives, designing with and for maker communities, peer-content creation, and crowdsourced design and innovation strategies. Furthermore, minimum viable product development strategies have underscored the benefits and gained versatility of extending the design into the use time. The constraints of and possibilities opened up by situated use are simply difficult to anticipate during concept design. In all these research and practice approaches, the key underlying questions are what and how are people capable of engaging with designs during use time and how is their design engagement supported and built upon.

In this chapter, we first go through what different disciplines have established about users' design engagements during use time. We then present a taxonomy of active use that reveals the impressive range in which active-use phenomena happens. Use as is, active use, locally new designs, and globally new innovations mark different intensities of engagement. These can concern the material form of design, new uses, new meanings, and adjustment to local settings. Equally, there are collective forms of active use that shape communities and organizations, ideologies and imaginaries, and global platforms that facilitate active use. All these aspects and gradations of user engagement can be found in hotspots such as digital-physical making activities, but many also in other digital and physical settings.

The chapter then proceeds to elaborate on how the design spaces can be temporally extended using the users' active design engagement through aging together design guideposts. The aging together methodology is illustrated with a case study of building an information infrastructure for a communal housing initiative with the Active Seniors association, a case that outlines how designs can evolve and blend with infrastructuring and maintenance of community resources.

Chapter 4: Design Participation in Steering Wide Sociotechnical Change

This chapter outlines what design participation can offer for steering sector-wide systemic changes. Because of the difficulty of transforming sector-wide patterns of production and consumption to more sustainable footing in key societal sectors such as transport, energy, and land use, transitions governance is an increasingly central area of concern among policy, business, and design practitioners, as well as among academic researchers. Society-wide changes hold long-time frames and wide scope, spanning regulation, technology development, altered consumer practices, taxation, and new business creation (and so on). The chapter begins by introducing the basic facets of why such society-wide sociotechnical change is so difficult: path dependency, vested interests, technical and infrastructural lock-ins, and sector-wide and cross-sectoral interlinkages between different path dependencies forming "sociotechnical regimes."

We then move to introduce the key transition governance approaches and how design participation can contribute to them. We elaborate these possibilities further through introducing the mid-range transition pathway toolkit (MTPT) and mid-range transition arenas. There, design participation fosters sector-wide changes by helping diverse actors in envisioning and building pathways toward a more sustainable future with hands-on tools and improved procedures. After briefly recounting the evolution of these tools to date, we shift gears and examine the process by which the MTPT was designed to exemplify those facets of intermediate designing that become foregrounded in codesigning for transition. The chapter ends by considering the potential as well as the reorientations that may be required from designers in catalyzing such transformative changes.

Chapter 5: Diversifying Participation with Mixes and Ecologies of Knowledge

An important subtext across the chapters has been that it is not always, even in principle, possible to identify and attract the "right" participants or even a "right mix" of participants. The differences among the people seen relevant for design may call for different arrangements, and shoehorning all the potential participants and reasons for design participation into any single "format" or method will do injustice to some participants as well as miss opportunities with others. Meaningful participation in different contexts, for different issues, and with different people is a practical (and often anxious) achievement across several competing "goods."

One avenue toward resolving this quandary is the concept of a "participation mix" in which different but related formats complement each other. Participation mixes can be built to offer complementary ways of design participation to enable meaningful participation across the differences present among potential participants; differences in the preferred ways of participating that these people hold; diversity in the design and knowledge needs present in the design project(s); differences in potential participants' design ability and willingness to engage in design; and often also (the finite) resources available for organizing participation.

The idea of a participation mix is not new per se, yet it has remained less elaborated than it should be. Medium and large participatory and human-centered design projects have always conducted complementary user research and user-involvement activities either in sequence or in parallel to address different users and knowledge and the design needs of the projects. Just as well, large private companies have hosted user groups to similar effects, segmenting them according to voluntary participants' abilities and interests in, for example, segments involved in future trend scanning, concept development, usability and user experience testing, end-user innovation, and prioritization of development requests. What has remained missing has been reflections of the rationales for building up the mix; thus, we take steps toward that in the end by examining the participation mix we developed for the Oodi Central Library in Helsinki.

In the latter part of this chapter, we discuss how the resulting complexity of how different participations and ways of knowing the citizen/user/customer insights relate to organizations and institutions has been treated superfluously. To remedy this, the chapter elaborates on "ecologies of user knowledge" mapping tools that help a more analytical approach to starting points for and long-term achievements resulting from design participation. The chapter consequently elaborates the import of ecologies of user knowledge with two cases at the Finnish National Broadcasting Company.

Chapter 6: Making Sense of Design Participation Methods Through Designer–User Configurations.

This chapter zooms out to give the advanced readers an overview of different approaches by which design participation is carried out. There has been dramatic growth in the ways of involving users in various product, service, space, and systems development activities. Various categorizations, listings, and mappings have been proposed for making sense of this user-involvement landscape. The chapter first introduces the main types of currently available clarificatory mappings used in design for, with, and by users, as well as the shortcomings associated with them. It then builds on this analysis to elaborate how such mappings could be better organized in the future, stressing the importance of distinguishing between techniques, methods, methodologies, and approaches. While sensible mappings are difficult if not impossible to ever achieve regarding techniques and methods due to their fluctuating status in design activities, wider approaches to designer–user relations configured in time and space appear to be amenable to meaningful differentiation.

The chapter consequently proceeds to elaborate a typology of nine approach-level user-involvement configurations based on resemblance relations and prototypicality between the methodologies within them. User-innovation communities, hybrid-innovation communities, (host) coordinated user design, design-in-use, collocated collaborative design, human-centered and UX design, designer immersion, and user inspiration are outlined regarding their predominant designer–user configuration. The final analyses in the chapter discuss the different positions that designers and users hold in each and use the designer–user configurations to elaborate how some key research areas within design participation—exemplified by user-innovation research, PD, and service design—factually span several approaches and thus, again, are likely not the best units to make sense of the user-involvement landscape and its evolution either per se. Yet the designer–user configurations help to better understand their different profiles and complementarities in between.

Chapter 7: Assessing Contexts for Scoping and Planning Design Participation

The book overall advances an argument that there is no single right way to pursue design participation. Rather, to make it real and not just the academic high ground, it must be adjusted to the contexts, aims, resources, and so on that are at play. Yet this does not mean that anything goes. On the contrary, one needs to take their bearings carefully to land in well-suited ways to carry out design participation. Assessing the contexts of design participation then becomes important, and the final chapter proceeds to provide a set of concepts and considerations for approximating it. Drawing from innovation studies and S&TS, the chapter outlines sociotechnical change and actor-group-related phenomena that typically need to be gauged upon considering whether and how design participation could be pursued.

These considerations begin from the characteristics of the issue and setting in which design participation is sought. Building on Chapter 4, the considerations move to assessment of salient sociotechnical structuring, such as presence and depth of path dependencies, the depth and width of infrastructural lock-in and interdependencies, and the size of sunk investments, paired with assessing the key actors in the domain, their characteristics and interrelations, and the arenas in which they interact and introduce one way to map them that is useful for approximating the context for potential design participation actions.

Not pretending to provide an exhaustive list of issues to be considered and insisting on the continuous translation between design participation choices, the chapter walks these considerations through three design participation cases in the Finnish energy transitions. The first one is a case of the as yet unrealized potential benefits of two actor groups, in this instance the use of ambient heat reserves to decarbonize district heating. The second concerns supporting an emerging phenomenon that is recognized as societally valuable but falling between the cracks of different actors: "hybrid heating" renewable solutions in detached houses in Finland. The final one is a case that seeks to better steer the evolution of the more structured and high investment industrial field of hydrogen development within advancing energy transitions.

Chapter 8: Conclusions and Notes About Positions Taken

The final chapter recaps the most salient arguments of the book and positions them further in the participation and design-related literatures.

CHAPTER 2

Design Participation Work and the Workshop as Its Microcosm

Introduction: From Methods and Cases to Pragmatics and Work

In this chapter,[1] we first further elaborate on a "deficit bias" in how design participation has been conducted and reported in prior research, emphasizing methods and illustrative cases. This has left the pragmatics and strategic alignment of design participation neglected in the academic and practitioner-oriented literatures alike. We then elaborate on six different kinds of design participation work to better conceptualize these neglected aspects that are factually needed to ever successfully realize design participation and exemplify how they play out in two quite different workshops arranged in two design participation projects.

The dominant premise for involving users in design has, ever since the 1970s, been an observed deficit in participation opportunities in design, planning, and technology development. This has contributed to a poor fit between

[1] Earlier versions of analyses that comprise this chapter can be found in the following four articles: (1) Hyysalo, Sampsa, Kohtala, Cindy, Helminen, Pia, Mäkinen, Samuli, Miettinen, Virve and Muurinen, Lotta. (2014) Collaborative futuring with and by makers. *CoDesign* 10(3–4), 209–228; (2) Kohtala, Cindy & Hyysalo, Sampsa. (2015) Anticipated environmental sustainability of personal fabrication. *Journal of Cleaner Production* (99), 333–344; (3) Botero, Andrea, Hyysalo, Sampsa, Kohtala, Cindy and Whalen, Jack. (2020). Getting participatory design done: From methods and choices to translation work across constituent domains. *International Journal of Design* 14(2), 17–34; and (4) Hyysalo, Sampsa, Greger, Sebastian, & Hatami, Zagros. (2013) Co-designing the future health care. In (eds. Kivelä, Kaisa, Tyyri-Pohjonen, Sanna and Kinnunen, Krista,). *Living + For Better Living Environments*. Helsinki: Aalto ARTS Books.

S. Hyysalo, *Design Participation*,
https://doi.org/10.1007/978-3-031-95437-5_2

humans and technologies and a lack of democratic governance regarding technological change resulting in social inequalities and environmental harms (e.g., Bjerkness et al. 1987; MacKenzie and Wajcman 1999; Greenbaum and Kyng 1991). Research and practice in participatory design and codesign has, in turn, focused on addressing this participation deficit through developing arrays of techniques, methods, and approaches to involve and collaborate with users. The net result has been an accumulation of hundreds of participatory and codesign methods that have been introduced and evaluated (e.g., Muller and Kuhn 1993; Muller and Druin 2012; Bødker et al. 2004; Curedale 2013; www.actioncatalogue.eu; peoplepowered.org; participedia.net).

Some technique and method development has taken place in response to analyses of the political economy and the development of design principles to meet this, as in the early "collective resource approach" (e.g., Bjerknes et al. 1987; Ehn and Kyng 1991; Kensing and Munk-Madsen 1993). Others have resulted from responding to project particularities and the needs of different participant groups (e.g., Mattelmäki 2006; Keinonen et al. 2008; Druin & Muller, 2009). The resulting research reports and grey literature outline successful projects and case studies in hopes of providing examples and inspiration for those who might seek to make up for the deficit. These two genres (single method and single case study) tend to intertwine so that project reports often motivate new methods and ways of working with users as the novelty pursued in a success case, making method introduction one of the most common ways to legitimize academic novelty.

But times they are a-changing. Given that there now are hundreds of available methods and inspiring cases, researchers and practitioners have been increasingly concerned about how one actually chooses, adapts, and modifies methods and project work to achieve successful outcomes. Lee et al. (2018) performed a cross-comparison of 13 codesign projects and concluded that designers had to always attend to local contexts and modify methods as situated practice. They further characterize recurring problem dimensions and contingencies, such as openness of the brief and distribution of power, that result in a set of related design choices. The adequacy of method-and-case-study orientation has been eloquently questioned by Woolrych et al. (2011) and Johnson et al. (2014a) by evoking the metaphor of cooking. While methods are presented as "recipes," much else required for competent cooking practices remains unattended to. Woolrych et al. (2011) stress paying due attention to the needed "ingredients," i.e., the low-level techniques and mundane acts that are needed in operationalizing the recipes—the methods. As importantly, they underscore that in real life few people make recipes but prepare "meals" amidst the contingencies introduced by time pressure, available ingredients, and preferences of the participants. Similarly, people pursuing design participation in industry, the public sector, or among activists seldom deploy methods (akin to what I might do as a university professor to see what comes out of a particular recipe) but adapt them to accomplish something as part of broader and more concrete design participation *work*. This is to say, while the method recipes, good case descriptions, and accompanying how-to descriptions have been pivotally important in guiding practitioners to successful implementation (e.g., Beyer and Holtzblatt 1998; Holtzblatt et al. 2005; Slesvik-Visser et al. 2005; Sanders and Stappers 2012; Bødker et al. 2004), they still fall short in addressing empirically and conceptually what is involved in the everyday deployment of design participation, particularly by people other than those who have developed a given method.

The recent wave of studies that have studied up close how design participation is accomplished in practice consistently show how people involved, particularly those seeking to invite and coordinate participation, engage in both mundane and strategic work that permeates how design processes become organized, how methods are deployed, what kind of design outputs follow—and eventually what is made out of those outputs at the end of the day (e.g., Jensen and Petersen 2016; Hyysalo and Hyysalo 2018). Bødker et al. (2017) similarly point to several layers of strategizing that are needed to sustain participatory design across moments, projects, and organizations and the associated backstage, unacknowledged work. Eriksen (2012) describes the materializing processes in staging and formatting codesign and the "intermediate designs"—that is, the tools, templates, settings, rules, and facilitation procedures involved (Eriksen 2012; Mattelmäki et al. 2011).

Hyysalo and Hyysalo (2018, p. 43) similarly phrase that "all the kinds of work that go into collaborative design, should be examined as co-constitutive to the processes, results, and further uptake of the collaborative design outcomes... [as] internal issues of stakeholder involvement and not just as external context or excludable routine execution". Mundane work refers to the variety of actions that range from arranging spaces for workshops to contacting participants to sorting outputs to guesstimating what the participants can get done in a given timeframe. Such actions could be seen as low-level design activities or as part of "silent design" in organizations (Gorb and Dumas 1988). Yet these could just as validly be seen as janitoring, recruiting, secretarial work, or, just as well in some cases qualitative data analysis related to design. Hyysalo and Hyysalo (2018) and Botero et al. (2020) argue that such mundane work permeates collaborative design and plays an important role in its outcomes just as much as the methods and principles that academic research has paid more focused attention to. Pinto and Nango (2025) and Vertiz Marquez (2024) demonstrate the centrality of a range of mundane work in design participation focused on resistance and social reproduction. Their works further underscore how design participation often serves *strategic work* within and beyond design,[2] in their cases furthering safe spaces for

[2] Recent participatory design work has described what we refer to as strategic work as "institutioning" (e.g., Hyubrechts et al. 2025). As we discuss below there are several albeit interlinked types of work involved and in our view these are not some separate

children to play in urban areas and aiding indigenous peoples' pursuits for recognition and identity negotiation. Just as well, strategic work in design participation can become a vehicle for organizational transformation or an avenue for marketing and public relations (Hyysalo and Hyysalo 2018; Body and Juninger 2008).

The net effect of mundane and strategic work has been captured well by Jensen and Petersen (2016) who analyzed how the project pragmatics and characteristic task series overrode and straddled the aims of user empowerment and fears of user exploitation in Danish user-driven innovation projects. In all, the realization of design participation outside sheltered academic settings has become ill-documented and analyzed, particularly regarding the full scope of citizen engagement in projects without focusing on just one or another narrow aspect within it (Steen 2011; Jensen and Petersen 2016).[3] This entails shifting focus from principles, ideals, and methods to the overall arrangements in which the participation takes place and through which the participants engage in the matters involved (Fig. 2.1) (Steen et al. 2011; Pirinen 2016; Hyysalo and Hyysalo 2018; Jensen and Petersen 2016).[4]

The types of work have been further clarified by Hyysalo et al.'s (2019) analysis of setting up a designer–user community. Taking their cue from symbolic interactionist studies of work (Strauss 1993; Clarke and Star 2007)

participatory design action or additional aspects to it but in fact always constitutive of design participation and at that permeates and qualifies all its other constituents (Hyysalo & Hyysalo, 2018; Hyysalo et al. 2019; Botero et al. 2020). One could say that institutioning is a pursuit that is realized through these types of work.

[3] An analogous critical analysis on participation in environmental governance and technology assessment has been put forward by Voß and Amelung (2016), Chilvers and Kearnes (2016), and Chilvers et al. (2018). See Chapter 1.

[4] One way to think about these interrelations further is to follow Irwin et al. (2013) and Botero et al. (2020) to see design participation involving a translation process, as it has been elaborated on in sociology of science and technology (e.g., Callon 1986, 1998; Latour 2005) between and among a variety of constituents, which actors perform in order to design collaboratively and get their ideas developed and implemented. Callon (1986; 1992) uses translation to describe processes in which human and non-human entities are transformed into new assemblages that have capacities to act in novel ways. Translation actively establishes relations between entities and materializes into arrangements that can sustain alliances and advance ideas and causes (Callon 1992, pp. 81–82). Actors who do translation must (re)define the identities of entities they seek to assemble (e.g., methods, mundane work, design, and strategy in Fig. 2.1) in relation to the others (within the range of choices open to them), and in this capacity, translation is never a process of just linking unfettered elements but transforms interplays and interdependencies between things or actors involved (Callon 1986; Star and Griesemer 1989). Translation reminds us that there is a variety of already existing elements, relations, and agendas that impact the work and the interplay between and among the constituents of design participation. This may imply misalignments: translations are never pure and involve compromising the origins and nature of elements assembled (Latour 1993, p. 253). It is important to recognize that translations in design participation are of a specific kind (cf. Irwin et al. 2013). Not all translation is design participation, nor does all work in design participation require translation: some tasks are performed without being qualified by the likes of strategic considerations.

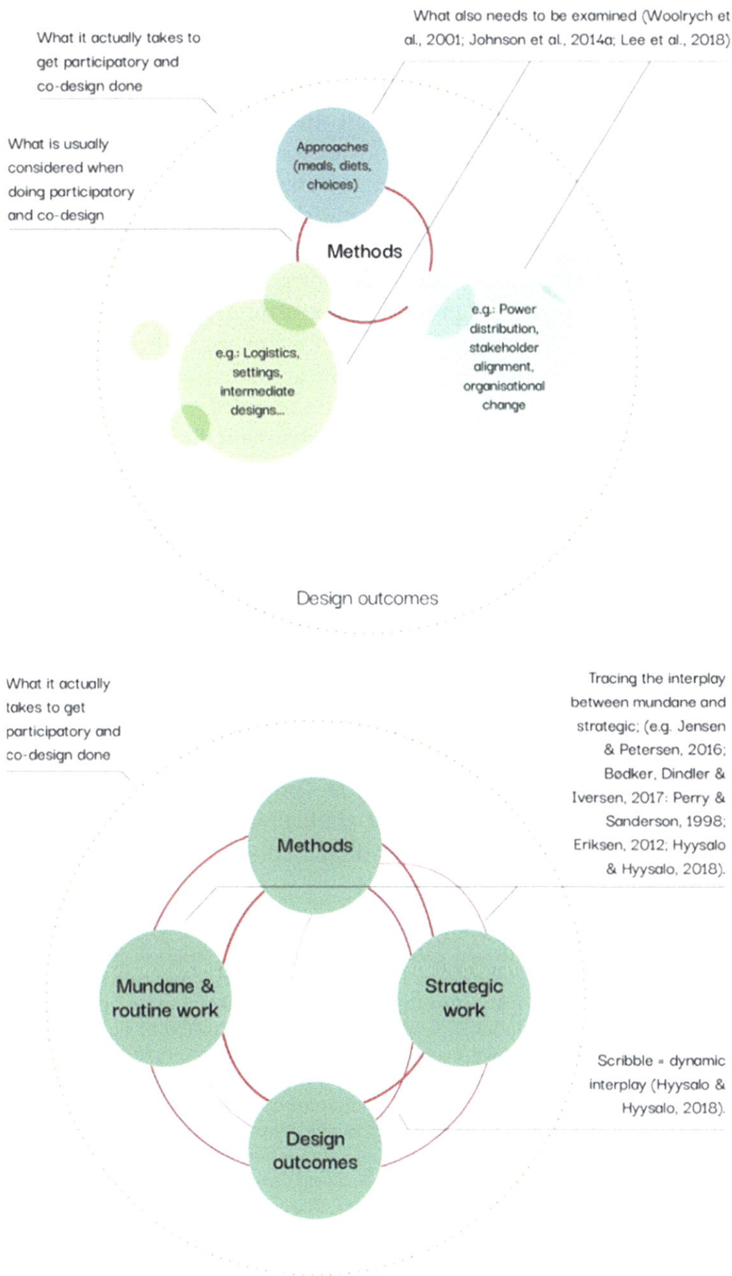

Fig. 2.1 Methods is only one domain of codesign and participatory design work, which is constituted by an interplay between mundane and strategic work and design outcomes, figures adapted from (Botero et al. 2020) with permission

and the way it has been used in analyzing participation in community innovation (Verheig et al. 2016), they detail six types of work involved.[5] Active *framing work* is required for delineating the issues for design participation and interlinking principled positions about the issues, the kind(s) of participation potentially involved, and the rationales for the participation and further turning these into concrete proposals. In the Hyysalo et al. (2019) case, framing work acknowledged complementarities between three different and partially conflicting ideals of design democracy: informing decision-making by elected officials with citizen views, empowering representative citizens in making design decisions, and providing opportunities for design-savvy citizens to directly affect the design of large public projects. Framing work does not need to, and often does not, result in a unified and shared consensus view among the key actors, nor does the framing necessarily identify a problem–solution pair that would then be instrumentally solved by the design participation. It concerns framing of the participation and its relationship to design, as much as the framing of the design problem(s) in view of its solution(s). Yet the remaining underlying tensions do affect later design phases, as they may need to become settled at a later time to accomplish the design participation.

Framing work is thus part of alignment work (Verhaegh et al. 2016), which also includes *relevance work*, that reaches toward both inward and outward actors. The inward reach concerns about how design participation connects to other aspects of advancing design and concrete targets for the design participation. Particularly in larger projects, the inwards' relevance work should not be belittled. There may not be just one single "early phase" that design participation could target but a series of gradually opening and closing design spaces for different spatial, service, and digital concepts (Botero 2013; Murto

[5] "Work" is here used in two interrelated senses. The first sense is the descriptive observation that what is involved in design participation is not just design or participation but a wider combination of different kinds of practical actions. Intertwinement of these actions is necessary for the design participation to carry through, yet, at the same time, the different types of work also hold independent existences. They require different skills to carry them out, and because of this, they may even be carried out by different people. Importantly, those skills and people are seldom in a one-off performance but deployed in other projects and settings as well, with similar or different interconnections. This is why the category of work is so central in participation practice: as the work is rehearsed time and again, it furnishes those carrying out the work with an anticipation of how other actors involved and implicated by the work are likely to respond and how those responses are patterned, allowing for skilled and responsible decisions and trade-offs between competing goods to participation (cf. Irwin et al. 2013). The second sense is that work as an analytical term denotes the interrelations between objects, subjects, engagements, and their context relationships. Work here points to the ongoing and unresolved character of practical action: none of this work "solves" the way participation is to be conducted for good but elaborates and transforms it in a given project and beyond it in other settings as the elaboration is not a one-time act of setting and solving the need for and the accomplishment of participation (on the nature of work, Star and Strauss 1999; Engeström 2000; on the persistence of the problems of participation, Irwin et al. 2013; Marres 2012, pp. 40–58; Kelty 2020). More on work as an analytical category and its ontic grounding are in a later footnote.

2017). Hyysalo et al. (2019) demonstrate how the often-ideal position of the project team running ahead of other planning resulted in considerable added burden, extra time, and increased ambiguities and uncertainties over arranging design participation and incorporation of its results. Just as importantly, outward reaching relevance work elaborates what might be there in the participation and for whom. In all thus "staging" participation intra and inter-organizationally with respect to other actors ongoing practices, processes, and interests (cf. Clausen et al. 2020).

Selection work concretizes framing and relevance work into seeking out, seeking to interest (cajoling), connecting to, and ultimately selecting actors who get to participate. Selection work can be a detached proceduralized operation of sending an invitation to a randomly chosen sample of the population and selecting from among the volunteers, as in some deliberative mini-public formation, yet then involving various balancing acts to retain the sought representatives in the sample among respondents' attributes. In such a selection, the matters and issues at hand do not affect the selection principles or process (Grönlund et al. 2010). Much of the design participation is however premised on giving the possibility of participation to affected and potentially affected people, which necessarily intertwines the relevant participants and the matters participated in, implying that the general population or even place-based or some socio-economic group is seldom the adequate backdrop (cf. Fung, 2006). Furthermore, as design participation does not just seek to elaborate a public opinion but also to transform the matters involved, the envisioned *format* of participation affects the *relevance* of the matter for the potentially affected, thus interested people who may or may not choose to participate (Marres 2012). Even further entanglements ensue in the common cases in which the actors who "host" participation do not already know what matters and issues are involved, but participation is sought for better understanding of what the matters and issues are, and in a perhaps disconcerting turn, there is irresolvable uncertainty as to who may become affected and who is capable of elaborating on and designing for the matter, before (some) participatory process has ran its course (Marres 2012; Hyysalo et al., 2014; Hannukainen et al., 2018). Selection work in design participation thus tends to involve a quite complex (pre)consideration about the *relations* between actors, matters, formats, and competencies (see below this chapter and Chapter 7).

Selection work is not specific to design participation, not even to public participation, as there is an ample amount of reaching out to and selecting people in many walks of life. Yet particularly in design participation, the selection work can require skill and experience in order to think through and anticipate the people who might be potentially affected, different people who could be contacted and could be interested in participating, and how to legitimize the choices in the process for the potential participants and other actors involved. Importantly, as is the case with all the other work types, the selection work cannot be performed in a social and material vacuum, and whatever formats, wordings, and procedures are used for reaching out to people and enticing them to participate will affect who eventually participates, as we see later in this chapter and in Chapters 4 and 5.

Selection work is typically further tied into and constrained by *competence building work* among the organizers—the amount of people available for organizing and their skills and experience in running the type of design participation that is being pursued. If the organizers have large numbers of competent people and time at their disposal for running design participation, selection questions can become solved by just adding the numbers of participats. Similarly, many relevance and framing questions may find settlement by expanding the scope and focus of participation. But such time and resource availability is seldom the case.

Intermediate design work and *collaborative design work* are likely the types of work people recognize most readily in design participation, be they designers or other organizers of participation. Intermediate designs are designs that do not carry into the final production but help the participation process to progress toward it (Erikssen 2012) or stage the interactions (Clausen et al. 2020). Examples of such designs are purposefully designed probe kits, tailor-made notational systems for work modeling, designs of workshops, design games, and models that are to push ideation further (e.g., Mattelmäki 2006; Vaajakallio 2012; Eriksen 2012; Erikseen et al. 2014; Clausen et al. 2020). Intermediate designs differ from means such as off-the-shelf prototyping software or stacks of Post-it notes, which are, as such, used as means or building blocks for reaching intermediate or outcome design. Intermediate designs also differ from the early versions of the outcome object, if the progressing prototype is not one of the key mediating means between users and designers (Botero 2013). This said, the borderlines between means, intermediate designs, and evolving final designs are blurry, particularly in cases where the outcome design is intentionally designed to allow for further design and evolution (Hartswood et al. 2002; Abel et al. 2011; Botero and Hyysalo 2013).[6]

[6] This points to the many forms that intermediate designs can take. At one extreme, one can merely assemble the building blocks—a table, chairs, paper, pen, and a stack of Post-it notes—and leave it up to participants to decide what to make of them. Common additions in codesign are making available a facilitator in the setting and relatively loosely structured templates to help organize the elements. The facilitator and the templates thus become an added *bricole* in what can be called *bricolaged intermediate design* (cf. Büscher et al. 2009). At another extreme, one may seek to build structured processes that hold a higher degree of assurance of some rather than other outcomes and some rather than other processual qualities. Such intermediate designs can be found in, for instance, simulation games and design games (Schrage 1999; Torvinen 1997; Vaajakallio 2012) and are well conceptualized as *gamified intermediate designs*. Such intermediate designs have a relatively tightly coupled script for action, which participants cannot meaningfully alter without altering the engagement with the intermediate design (Latour/Johnson 1987). Yet another end of intermediate designs lies in user-appropriated (meta-)design as making tools for design, where the structure of elements and their interrelations is closely defined. The meta-design and principles for using it are then elaborated to an extent that competent

Collaborative design work, in turn, consists of actions through which the proverbial participation rubber meets the road. This includes instructing, facilitating, time-keeping, recording, and elaborating on participant ideas and concepts and in general, making the participation run and proceed. It is noteworthy that other design participation work will permeate the situated achievement of participation or, in an oft-used phrase, "inherit" whatever was done and left undone before it. Surely, collaborative design work holds considerable capacity in accommodating other permeating conditions—but it does not have a limitless capacity to do that by any means. In Hyysalo et al.'s (2019) example, documentation and facilitation hung precariously close to collapsing, as all the other considerations had bestowed notetakers and facilitators with too many roles against the attainable training with the complexity of handling four parallel mixed groups with a unified pacing of collaborative design actions (see Chapter 5).

Finally, it is important to recognize that design participation is never a simple input–output affair. Its *outcome work* intertwines with relevance and selection work and further on with what kinds of devices—templates, design games, sticky notes, townhall meetings—are used to foster and capture participation in intermediate design work. It is acutely present in collaborative design work where a usual tension line runs between making space for free deliberation on one hand and producing communicable ideas and concepts on the other hand. The coordinators of participation, and often also participants, then engage in further elaborating on these outputs and, again tying onto relevance work, their potential for the host organization (or other actors seen as relevant to act on them). Curiously, the tangible results of participation may, in the process of communication to external parties, become overshadowed by them having emerged from a participatory process (or idea of that process) and vice versa.[7]

users can (or at least should be able to) pursue further design independently (Fischer and Gicciardi 2006). The design process of such *intermediate meta design* in this sense comes close to generative systems of representation and familiar design from various building block and toolset kits such as Lego that feature only suggestive goals and solutions.

[7] What is the import of "work" as an analytical category here? Strauss (1993) and Strauss and Star (1999) elaborate well on how work as a term for understanding socio-material relations directs attention to how things are getting done in their complex interrelations. While many other terms such as action, practice, activity, and assemblage do the same just as well, work has its own affiliations. Attending to work necessarily means traversing from the minutiae of work performances to those being particular enactments of skills, materials, and images developed over time, which are, in turn, occasioned in organizational and institutional settings and intersecting political economies and ecologies and the actors and arenas formed between them (Clarke and Star 2003; Engeström 2000). As a conceptual register work thus pushes our attention beyond any one layer of how things get done, inviting us to remain observant of both the specifics in their enactment as well as the historically wider constituents that are being enacted in any one occasion of work performance (Akera 2007). Epistemics tie in ontics, and as an ontic register work resides in a (left-)Hegelian line that runs across (neo)Marxist and pragmatist thought that places praxis and action as the prime motor that gives rise to gradual delineation of boundaries that then result in whatever became referred to as (more fixed) entities (Star 2006;

In summary, the types of work—framing work, relevance work, selection work, consistency of building work, intermediate design work, collaboration work, and outcome work—concretize what the permeation by strategic and mundane work entails. We shall next move onto elaborating on what these considerations hold in detail for running variations of the most common design participation format, a collocated collaborative design workshop.

Planning a Workshop as a Microcosm of Considerations: How to Define a Future Maker Space

The first workshop we discuss exemplifies a demanding workshop in terms of goals, the need to succeed, and the needs to organize the design participation. What we thus see is careful and minutiae explication regarding the workshop even in contrast to the second workshop we examine after this. The context for the 2012 workshop lies in seeking to define what should be in a library-housed maker space 10 years down the line at the new Oodi flagship library in Helsinki. In Oodi's preplanning phase, maker facilities and its space reservations were identified early as an area needing focused attention and insight beyond the library planners (see more on the overall mix of participation efforts in Chapter 5). The challenge then became how to best prepare for a maker space in a library setting that was to be opened several years down the line. Acquainting to a new area via books, expert interviews, and future procedures, such as trend extrapolation and scenario building, is a common practice in library planning (e.g., Staley et al. 2012). Recognizing the rapidly evolving nature of making practices, however, made it difficult for library planners or even high-level experts to adequately keep track of specifics of how the grassroots activities would be changing. To understand this, let us recount what maker space activities are comprised of.

Maker spaces offer access to low-cost digital fabrication equipment and electronics prototyping. They have become increasingly common in the post-industrial cityscape (van Abel et al. 2011; Nascimento and Pólvora 2013; Ramella and Manzo 2018). Although "making" builds on a tradition of handicraft and "DIY" (do-it-yourself), it is mainly associated with the use of digitally enabled, small-scale manufacturing tools in hands-on fabrication of material

Engeström 1987; Abbot 1995). In this fundamentally relational and temporal ontology, it is the living praxis that gives rise to subjects and objects and renders them perpetually evolving through the mutual resistance and suffering that at once shapes and retains them in relation to each other (Lave and Chaicklin 1993; Cole 1996). While relational ontology has been most prominently advocated by the actor network theory and its many descendant variants (Lezaun and Woolgar 2013), the relational ontic premises are more widely spread as Clarke (2005), Star (2006), Engeström (1996), and Latour (1996a, b) cogently observe. At this, the stress on work found in symbolic interactionist and activity theory orientations pushes analysts toward somehow accounting for the layering and structuring of sociotechnical relations beyond the most salient situated relationalities (Akera 2007). More on these themes in Chapters 5 and 7.

artifacts by users themselves, including electronics and physical computing experiments, design and engineering prototypes, and so on. While in name the "maker movement" was much driven by California-based Make Media and MAKE magazine and in practice through "Maker Faires" and open-access maker spaces including Fab Labs and hacker spaces from the beginning of the 2000s, it is a movement that nevertheless builds on variegated traditions of shared community technology workshops from past decades (Sivek 2011; Kohtala 2016; Halbinger 2018; Kohtala et al. 2020; Hepp and Schmitz 2022). Shared maker spaces typically contain milling machines for making circuits or casting molds, vinyl cutters, electronics workstations for microprocessor programming and electronics project prototyping, desktop 3D printers, and laser cutters. Product designs (often shared digitally) are realized by the users themselves and, due to their digital form, are created by users (designers and not design-trained people) and can be designed together with peers in other locations and shared as digital files that can be modified, often in large online repositories similar to practices in free and open-source software development. The number of maker spaces worldwide has grown rapidly into thousands since around 2010, while there is considerable variation in who uses maker spaces, from students in university Fab Labs to entrepreneurs to hobbyists (e.g., Eychenne 2012). In addition, new DIY strands are exploring areas, such as citizen science and urban agriculture, rural innovation and agroecology, material innovation, and "DIYbio" or biodesign, and the like activities and are conducted in their own communities and spaces or included in the repertoire of already established maker spaces.

Framing Work for the Future of Sustainable Maker Spaces

All this makes making a fast-moving target. Technologies, practices, and maker communities evolve rapidly with new platforms for design, sharing, and collaboration as well as equipment for distributed manufacturing. New areas such as biohacking and pharmahacking are expanding along with a renaissance of more traditional crafts. Manifestos are many, and the outlook of the future of making has continued to depend on who assessed it: owing much to the evolving and also quite diffuse character of making activities. Yet in a major library project with a new building, it was necessarily tied in to a slow planning process in which the requirements for future library maker spaces would somehow still need to be anticipated several years ahead.

Against this backdrop, the best experts to gauge the future of making likely would be those people who were at the forefront of different making practices. Consequently, collaborative forms of futuring thus appeared sensible as the knowledge needed was likely concentrated in active makers. The usual "participatory futuring" methods such as Delphis, charrettes, or future search conferences were however all geared toward broad future development (Glenn and Gordon 2009), which would still leave solution information unaddressed.

In contrast, lead-user workshops (Churchill et al. 2009) held a higher potential for identifying both trends and solution information, as lead users typically face future needs in domain years before most people do (von Hippel 2005). This consideration, paired with Aalto University's INUSE research group's interest in trialing advanced forms of design participation in the Finnish public sector, led to a proposal to conduct a workshop for the maker spaces of the central library in 2012, together with the Oodi planners and Aalto University's researchers.

Relevance Work: Ensuring Long- and Short-Haul Relevance

The workshop preparations were conducted as a collaborative effort between the Oodi participation planners and four Aalto university researchers. Aalto's researchers had run lead-user workshops before (Hyysalo et al. 2015), and this experience suggested that rather than aiming at just one workshop, ongoing interactions between library planners and lead users could be beneficial to arrange, particularly if this could be anchored to a concrete iterative onsite development (Hartswood et al. 2002; Büscher et al. 2009; Botero and Hyysalo 2013). At this, the team assessed that it could also be beneficial to seek to establish links to Finnish maker communities and extend the number of users invited. Furthermore, the planned maker facilities in a public setting would present an opportunity to consider and promote the sustainability of future-making practices to harness them for public social and environmental good rather than for the production of yet another set of unrecyclable hand-waving plastic Yoda figurines.[8] Such sustainability assessment of current and future maker practices was part of the dissertation project of one of the design researchers, and it added to the potential benefits of collaboration between Oodi planners and Aalto researchers (eventually published in Kohtala and Hyysalo 2015; Kohtala 2016).

Meanwhile, it was known to the design team that southern Finland, Helsinki in particular, had several different types of maker communities and independent professionals and semi-professionals who were highly proficient in making. The situation was not atypical to other European localities, but it did present an interesting terrain for anticipating what the Finnish makers as well as other citizens might require and prefer from a public maker space in 2020. The southern Finland maker scene was also very international: alongside native Finns, it featured people from several European countries and the Americas, many being internationally well networked.

This cast several parameters to the workshop design to be realized:

- Gain in-depth understanding of the key trends that will affect making practices, equipment, and spaces 7–10 years down the line.

[8] Various Yoda figurines were the most common 3D printed, and discarded, item globally for years, as they stood for being a geek—more on this in Chapter 3.

- Identify concrete solutions and no-go issues regarding planning reservations to a maker space to be inaugurated in 7–10 years' time.
- Be better able to elaborate sustainability aspects of present future-making practices and to support more sustainable making in publicly funded settings.
- Initiate interchange between the library planners and local maker communities, at best resulting in some form of exploratory co-ownership of the pilot maker facility to be opened in a few months' time.

This assessment was presented to the Oodi library planning steering group who agreed that it made sense, and thus, a joint effort to develop a participatory event to aid the design of maker spaces was agreed on.

Selection Work

These parameters and aims for the workshop emphasize that participants would need to hold in-depth knowledge of present and future making and, as a group, widely cover the different facets of maker culture. Thus, most people would not be able to meaningfully participate. If we stopped somebody on the street and asked, "What do you see as the most important trends in digital and sustainable making in 10 years' time?", the answer would likely be just a blank stare. Even after the person would have been provided with as good an information package as we organizers could prepare to aid their deliberation (cf. Renn et al. 1995), the person would have a very hard time explaining a view that could reach beyond what the organizers already knew. The same would apply to most people currently using maker spaces, as the knowledge we would need is a very specific kind, born only from years of active engagement with maker cultures (cf. Chapter 1).

This fundamental requirement renders moot the most common ideas about whom to contact for participation. Gender, age, ethnicity, education, income, domicile, acquaintance of the library, or balance among these in the group would not be the prime "differences that make a difference" (Bateson 1969/2000)—one could have just as meaningless an exercise with any and all generic parameters used as the key heuristic for the selection. In this context, we simply would not reach the design participation aims by involving a somehow representative group of present and future users. Consequently, representativeness of participants necessarily became a secondary, not primary, criterion in the participant selection process (see Chapter 1 on design ability and investment in participation). And there are people who meet the primary criteria. These are people who have such in-depth knowledge about future practices and solutions, sometimes characterized as "already living in the future of others," and are called lead users (von Hippel 1986, 2005). Lead users have encountered the problems and solutions related to the domain area far before others have, typically because of high interest, expenditure of time, and importance of the matter to them (von Hippel 1986, 2016). While this sounds appealing, there are no generalist lead users who could be readily contacted for participation. People become lead users with respect to specific domains and often only related to a few of its aspects (cf. Hyysalo et al. 2013; von Hippel 2016). This entails that these extremely useful and knowledgeable participants are rare—typically less than 0.1% of the population related to a given matter (von Hippel 1985; Churchill et al. 2009).

The downside of the lead-user method lies in the time-consuming task of identifying the lead users. To mitigate this, we decided to limit the search for participants to the pool of maker experts and activists living in Finland who would be "lead-user enough" and more likely to develop a long-term relationship with the library planning process. This choice was supported by one member of the project team (Cindy Kohtala) being embedded in the networks in Finland's DIY maker scene, and her decade-long membership added considerable legitimacy in furthering a sense of a common cause among the various maker communities.

The potential participants were identified by first determining the aspects of making and maker communities to engage to adequately represent the present and future of digital fabrication and maker spaces. The sectors of commercial, academic, third-sector, and local authorities were further sub-divided into fields such as ICT, engineering, digital fabrication, "hacking," "crafts," and "support organizations." This resulted in a list of 32 people, among whom the competences sought after for the workshop would be held by at least two invited individuals. The workshop date suited 13 participants, who upon further inspection presented a balance of male and female and most importantly represented all the competencies desired. Taken together, they held wide and deep knowledge on different facets of digital fabrication, shared workshops, open innovation, and peer-to-peer dynamics, as well as experience in organizing and facilitating participatory events. The participants were of four nationalities. Seven of them were active in international networks related to maker culture—that is, known for their international keynote talks they had given on making and open design and crafts-related topics (Fig. 2.2).

Constituency Building Work

This workshop and the envisioned possible collaboration with the pilot maker facility were made possible by uniting the interests of two parties, the Oodi library planning group and Aalto University's interest group, to study how leading makers envisioned the sustainability aspects of future making. Six people from the Oodi and pilot maker facility planning team observed and the lead planner gave a short presentation during the workshop, and the Aalto University team of four facilitated and documented the workshop. This combination of competencies and people power provided good grounds for designing ambitious formats for the workshop, as we could rely on sufficient resources available for preparing, running, and analyzing it.

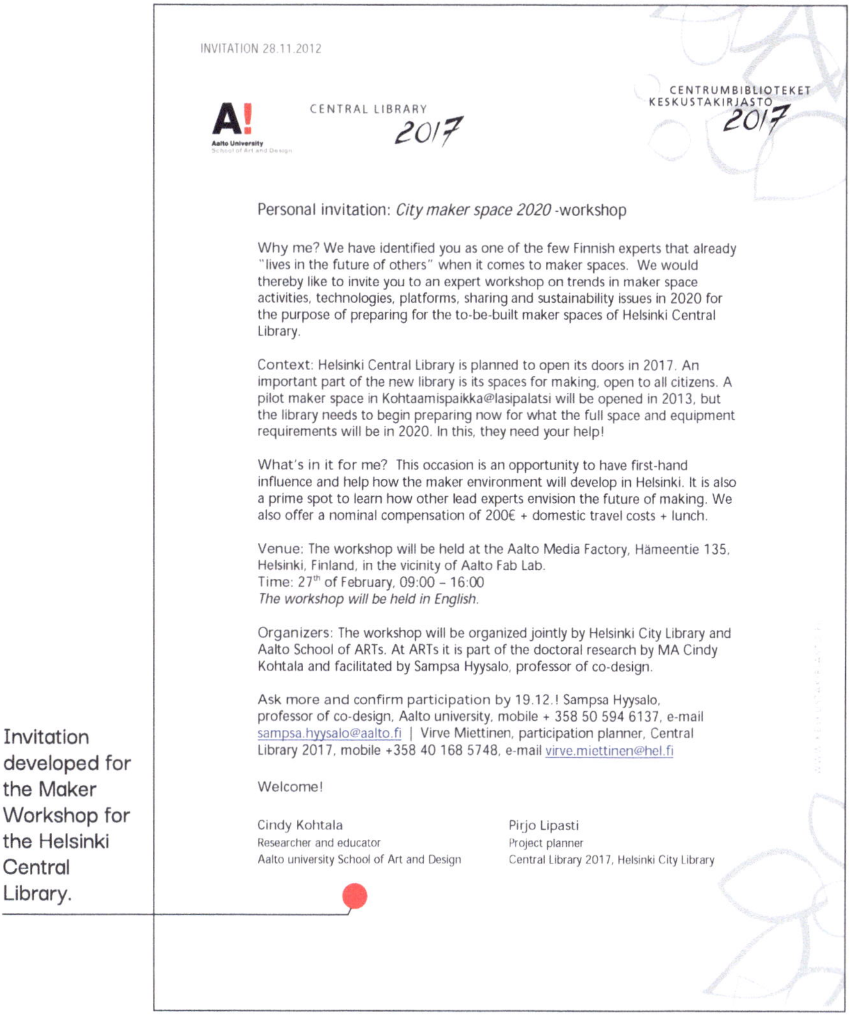

INVITATION 28.11.2012

A! Aalto University School of Art and Design

CENTRAL LIBRARY 2017

CENTRUMBIBLIOTEKET KESKUSTAKIRJASTO 2017

Personal invitation: *City maker space 2020* -workshop

Why me? We have identified you as one of the few Finnish experts that already "lives in the future of others" when it comes to maker spaces. We would thereby like to invite you to an expert workshop on trends in maker space activities, technologies, platforms, sharing and sustainability issues in 2020 for the purpose of preparing for the to-be-built maker spaces of Helsinki Central Library.

Context: Helsinki Central Library is planned to open its doors in 2017. An important part of the new library is its spaces for making, open to all citizens. A pilot maker space in Kohtaamispaikka@lasipalatsi will be opened in 2013, but the library needs to begin preparing now for what the full space and equipment requirements will be in 2020. In this, they need your help!

What's in it for me? This occasion is an opportunity to have first-hand influence and help how the maker environment will develop in Helsinki. It is also a prime spot to learn how other lead experts envision the future of making. We also offer a nominal compensation of 200€ + domestic travel costs + lunch.

Venue: The workshop will be held at the Aalto Media Factory, Hämeentie 135, Helsinki, Finland, in the vicinity of Aalto Fab Lab.
Time: 27th of February, 09:00 – 16:00
The workshop will be held in English.

Organizers: The workshop will be organized jointly by Helsinki City Library and Aalto School of ARTs. At ARTs it is part of the doctoral research by MA Cindy Kohtala and facilitated by Sampsa Hyysalo, professor of co-design.

Ask more and confirm participation by 19.12.! Sampsa Hyysalo, professor of co-design, Aalto university, mobile + 358 50 594 6137, e-mail sampsa.hyysalo@aalto.fi | Virve Miettinen, participation planner, Central Library 2017, mobile +358 40 168 5748, e-mail virve.miettinen@hel.fi

Welcome!

Cindy Kohtala
Researcher and educator
Aalto university School of Art and Design

Pirjo Lipasti
Project planner
Central Library 2017, Helsinki City Library

Fig. 2.2 Invitation to the maker space workshop

Intermediate Design Work: Methods, Templates, and Timing

The usual form of lead-user workshop follows a process outlined in von Hippel (1986) that departs from identifying important trends and needs, then identifying lead users regarding these, and finally working with the lead users in a workshop to generate product/service concepts (Churchill et al. 2009; Urban and von Hippel 1988; Lüthje and Herstatt 2004). The first part of the lead-user workshop focuses on trend exploration by lead users, explaining the challenges and trends they are facing. In the second part, the lead users and organizers together concretize solutions to meet these trends (Herstatt and von Hippel 1992; Churchill et al. 2009).

To enhance our participants' capacity to elaborate the future trends, we decided to provide "double stimuli" or "scaffoldings" (Cole 1996; Nardi 1996). First, drawing on representing work and helping ideation through prefilled cards (Muller 1993; Halskov and Dalsgaard 2006), we decided to give the participants cards in six categories to be used in both trend identification and solution proposals. The pre-categorized cards would also allow the organizers to rely on participants' own coding of the issues they expressed (Fig. 2.3). Using the same categories throughout the day for trends and solutions eased the learning burden from the participants in between the trend and solution part and allowed comparison between the two (Hyysalo et al. 2014; Kohtala and Hyysalo 2015).

Second, for the solution identification part of the workshop, we turned to ideas of "full-scale participatory modeling" (Hornyánszky Dalholm 1998) to help people achieve a "hands-on future" (Ehn and Kyng 1991). The participants were to use a similarly sized Aalto University maker space as a proxy for the city library maker space in 2020 and were asked to place the post-it based cards directly onto the machines and surfaces to show what would be different and what needed to be considered for 2020. The rationale was that a rich contemporary setup could both aid and ground imagination (Ehn and Kyng 1991; Büscher et al. 2009).

Our workshop was to proceed through four phases. After all 13 participants introduced themselves, they were asked to write individually what would be the important trends in making and maker spaces in the next eight years (year 2020). The pre-coded Post-it notes featured five categories, "technology," "activities," "sharing/organizing/IPR," "safety/risks," and "other." These categories were determined based on our prior research (reported in

Fig. 2.3 Trend identification in the workshop: working alone, presenting to others, starring, and elaborating

Kohtala 2016; Kohtala and Bosque 2014; Kohtala and Hyysalo 2015), and we hoped they would ensure that the participants remembered to include not only equipment but many aspects of making and its future.

The second phase focused on the sustainability aspects of future-making trends, as this had been a grossly neglected area for the library services. Here we decided to use the "World Cafe" technique (Brown 2002), where the most heavily starred issues were moved to three flipcharts and the participants moved between these stations in groups to discuss the sustainability implications of each (using "sustainability" category cards).

The third part of the workshop was to be held in the Aalto Fab Lab. The introduction of the new library's concept was to be followed by asking the participants to proceed with the full scale participatory modelling through adding notes directly onto the machines and surfaces regarding solutions, using the same pre-coded Post-it notes as in the morning part of the workshop (Fig. 2.4).

Because of our previous experience of lead users tending to be extremely savvy and efficient in working on their own, it was only for the final fourth part of the workshop that we shifted the participants to collaborative group work. Three groups were formed to outline the activities, technologies, and outreach of the pilot maker facilities that were to be launched just months after the workshop. The participants were given the pilot maker space floor plans and were asked to elaborate directly on those what should be there, followed by cross-group discussion (Fig. 2.5).

As the day was tightly paced and very much "a one-time opportunity" that could not realistically be repeated or done again, timings between and within the phases were estimated carefully beforehand at five-minute intervals. Figure 2.6 is part of one of the internal runs through schematics that we iterated several times over. Accompanying the intermediate designing were parts and materials lists with initials of who does what and a checklist with deadlines to ensure nothing was left unattended by the time it might be too late to remedy it by other means. In contrast to many other events and productions, design participation cannot be fully rehearsed, just proxied, and piloted with

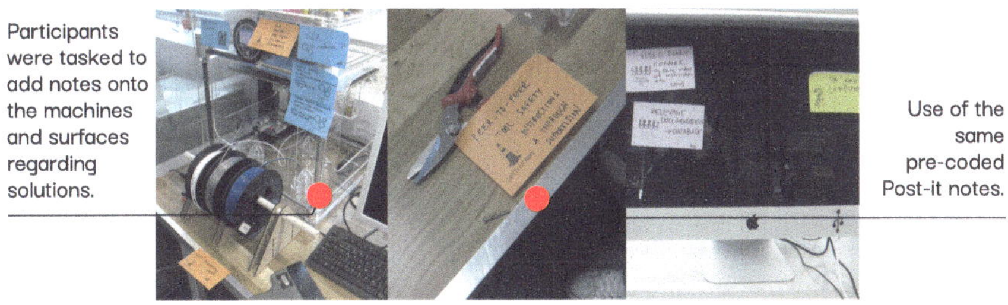

Fig. 2.4 Examples of solutions posted directly on the surfaces of the Fab Lab

Fig. 2.5 Concretizing the equipment and layout of the pilot maker space directly onto its floor plans in small groups

stand-in participants, and for this reason we did the intermediate designing in the very spaces we would be using in the workshop so that we would see clearly what the space layout would allow and restrict and could play-act some of the most critical moments.

Collaboration Work: Workshop Facilitation, Results Analysis, and Refinement

An important issue with intermediate designing for design participation is that participation cannot be over-defined (for more on this, see Chapter 4); hence, flexibility, improvisation, and back-up measures tend to be needed. For this maker workshop, we had a questionnaire about maker trends to help with the first part with the instruction that participants could turn to the questionnaire sheet and answer the statements there at once if they felt at all stuck with elaborating the trends on their own. Surprisingly, nearly all participants turned to the questionnaire only at the very end of the first part (i.e., did not need it), and filling it out resulted in only a handful of new trends on Post-it notes written by the participants.

In all, the exercises proceeded with little need for encouragement or facilitation; the participants seemed to have a zeal to spell out their ideas. This was by and large what we had expected from the participants with considerable lead-user characteristics, and this expectation was the basis for having the inspiration cards as unambiguous and "down-designed" as possible (i.e., simple graphics and text as category reminders). The participants produced 189 trend items in total. Each participant shared with the others the three most important trends they had written down. The top three trends were mounted on a wall, which was followed by an exercise of all participants starring which of the trends they felt were most important (not their own).

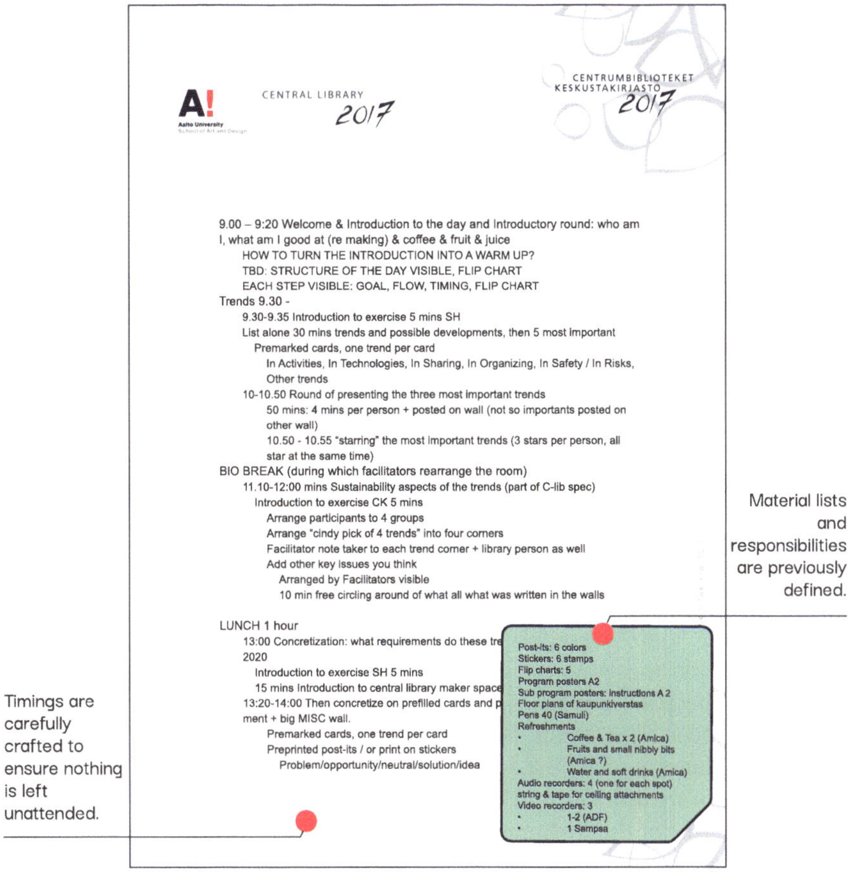
CENTRAL LIBRARY 2017

CENTRUMBIBLIOTEKET KESKUSTAKIRJASTO 2017

9.00 – 9:20 Welcome & Introduction to the day and Introductory round: who am I, what am I good at (re making) & coffee & fruit & juice
HOW TO TURN THE INTRODUCTION INTO A WARM UP?
TBD: STRUCTURE OF THE DAY VISIBLE, FLIP CHART
EACH STEP VISIBLE: GOAL, FLOW, TIMING, FLIP CHART
Trends 9.30 -
9.30-9.35 Introduction to exercise 5 mins SH
List alone 30 mins trends and possible developments, then 5 most important
Premarked cards, one trend per card
In Activities, In Technologies, In Sharing, In Organizing, In Safety / In Risks, Other trends
10-10.50 Round of presenting the three most important trends
50 mins: 4 mins per person + posted on wall (not so importants posted on other wall)
10.50 - 10.55 "starring" the most important trends (3 stars per person, all star at the same time)
BIO BREAK (during which facilitators rearrange the room)
11.10-12:00 mins Sustainability aspects of the trends (part of C-lib spec)
Introduction to exercise CK 5 mins
Arrange participants to 4 groups
Arrange "cindy pick of 4 trends" into four corners
Facilitator note taker to each trend corner + library person as well
Add other key issues you think
Arranged by Facilitators visible
10 min free circling around of what all what was written in the walls

LUNCH 1 hour
13:00 Concretization: what requirements do these tr
2020
Introduction to exercise SH 5 mins
15 mins Introduction to central library maker space
13:20-14:00 Then concretize on prefilled cards and p
ment + big MISC wall.
Premarked cards, one trend per card
Preprinted post-its / or print on stickers
Problem/opportunity/neutral/solution/idea

Post-its: 6 colors
Stickers: 6 stamps
Flip charts: 5
Program posters A2
Sub program posters: Instructions A 2
Floor plans of kaupunkiverstas
Pens 40 (Samuli)
Refreshments
- Coffee & Tea x 2 (Amica)
- Fruits and small nibbly bits (Amica ?)
- Water and soft drinks (Amica)
Audio recorders: 4 (one for each spot)
string & tape for ceiling attachments
Video recorders: 3
- 1-2 (ADF)
- 1 Sampsa

Fig. 2.6 The timing and to-do lists for the workshop day

The solution phase proceeded as planned for 90 minutes and produced 307 solutions or requirement ideas. The only facilitation interruption occurred at the end of the Fab Lab part of the exercise, when we asked the participants to concentrate on technology issues for a moment, followed by a similar request to focus on sustainability issues (Fig. 2.7).

In the final part of the workshop, the participants designed the layout of the pilot maker facility in the floorplan, in effect solving the library planners' internal debate over what equipment should be in the space and how to best place it. Second, the participants suggested that the library lend the space for pre-arranged co-hosting, an idea the library planners themselves had previously made successful for participatory events and music gigs but which they had no

Fig. 2.7 Generating solutions on equipment and surfaces

way of proposing to the makers previously. Third, the makers elaborated on outreach ideas for the pilot space and, after several imaginative marketing and campaigning ideas, converged on the idea that the best outreach would, in fact, also come through active co-hosting with various maker communities.

Outcome Work and Outcomes

The intermediate design of the workshop was geared toward producing several types of data for the library planning as well as for academic analyses. To this end, four audio recorders and two video recorders were arranged to cover most talk and physical interaction. As there were 13 people, there would likely be parallel actions and talk sequences, and thus, full transcription of the recordings could become unfeasible. The audio and video data were hence planned to be used as a back-up repository for the less intensive documentation methods from the onset.

The next layer of the documentation was photographs taken by both facilitators and library personnel. Altogether 691 photographs recorded every note written and the order at which they emerged. They provided a still picture trace of the flow of the workshop process. The written cards, 496 in total, were the next, and eventually the most directly useful, layer of the outcomes of the process. In addition, four facilitators made field notes after the day to record their observations about the workshop.

The data analysis began by first photographing all the cards in their locations to enable reconstructing the outcomes if needed. After this, the markings on the Post-it notes, categories therein, number of stars, author of each marking and their physical placement, and sequence in the events were tabulated. This, in turn, allowed us to present the materials to library planners, who went through all trends and solutions and marked which ones were immediately relevant, which ones applied to future planning, which ones were old news, and which ones confirmed matters that were yet uncertain for them (see Table 2.1). This produced an assessment of the yield of the workshop contents for their own planning work. For our academic analyses, we proceeded to calculate the distributions and trend-solution pairings of the issues raised in the workshop to see if there were differences between trend and solution information and information gained in different aspects of making.[9]

When we compared the trends and solutions regarding what was news, what questioned planners' assumptions, what supported their existing knowledge, and what was just nice to know, the relative amounts of "nice to know"

Table 2.1 Examples of solutions and their scoring by library planners

Category	*Post-It Contents*	*Location of post-it notes (in the Fab Lab)*	*Value for Library (Q = questions prev. ideas, S = supports ideas, N = new issue, NTK = nice to know*	*When? Future = 2020 Today = 2013*
Technology	Design the extraction of the fumes from the start, don't add it later	Laser cutter	S	F
	Choose low-cost accessible technologies that will also be found more easily elsewhere	Misc	S	T
Activities	Personal ["scout's"] badge in which skills, experience etc. Are collected—add item when [able to] use new tool	Large milling machine	N	T
	Access 24/7/365	Glass wall next to exit	S	F
Sharing Organizing	Hierarchy of good-bad materials on display (critical material thinking)	Bookshelf/ display case	N	T
IPR	"Video diary" corner: save videos of instructions etc	Soldering stations	N	T
Safety risks	Culturally independent [i.e., Neutral] symbols for security or safety functions	Large milling machine	NTK	T
	Ergonomics: lighting, posture, work time	3D printer, computer space	S	T
Other	Everything on wheels	Toolboxes	S	T
	Create user stories that help explain what any one of us can do in the space	Misc	N	T
Sustainability	Site-specific energy information/data	Misc	N	T
	Grouping jobs together; less waste, less energy, less time	Misc	N	T

[9] In another line of inquiry, we assessed the implications that the participants saw for the trends and solutions for environmental sustainability (Kohtala and Hyysalo 2015).

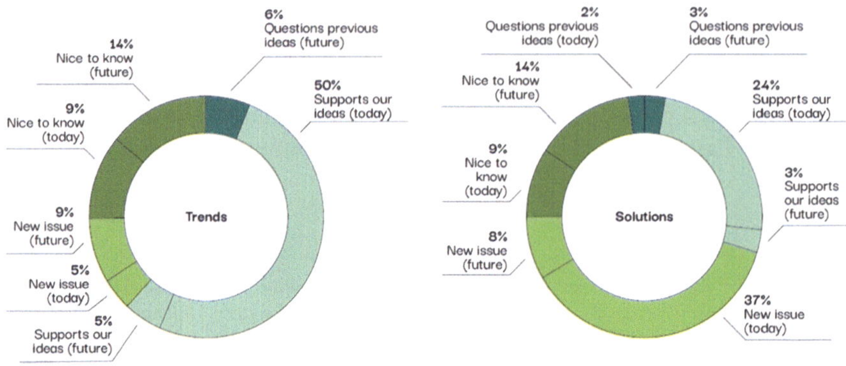

Fig. 2.8 Relevant scoring by library planners of trends (left) and solutions (right)

items were close to identical (25% and 23% for trends and solutions, respectively), and the same went for items that questioned extant ideas (6% and 5%), albeit with some difference with timing (i.e., now versus future). In contrast, the relative amounts differed greatly regarding the items that supported and those that were new to library planners (in trends, 55% supporting and 14% new; in solutions, 27% supporting and 45% new) (Fig. 2.8). Moreover, a great difference lay in the amount of supporting and new items that were relevant already in the planning of the pilot maker facility—ideas that already could be reacted to now.

Breaking these distributions down according to the categories revealed where these differences were accentuated. In technology trends and solutions, the share of new ideas in trends (24%) compared to the solutions (48%) was double (200%). This could be expected, as the technical details of making could arguably be challenging for library planners. But the ratio of novel solutions was triple in sharing, organizing, and IPR (339%), safety/risks (283%), and "other" (307%), and in activities, where the share jumped from just 3% to 50%, there was a 16-fold (1,667%) increase. This indicated that it had been relatively easier for library planners to grasp the overall development trends in maker technologies than to concretize what they would entail in everyday practice, highlighting the unique yield of design participation approach.

Reflections on the Library Maker Workshop

In 2014, a year after the workshop, the library planners told us that the workshop had kick-started a dramatic increase in their competency in maker spaces and provided access to a network of knowledgeable people who were willing to answer questions and assist. Over half of the participants in the workshop had stayed in active contact, and three different maker groups had appropriated the pilot maker space for their activities, both for open and closed events.

Drawing from user innovation research and participatory design, we sought a democratic approach to futuring and mid-range planning with deeply immersed makers, supported by PD techniques to aid the trend and solution identification. The coverage, depth, and concreteness of different aspects of making (not only technologies but also activities, sharing, IPR, and so on) would have been difficult to glean from working with experts or with more traditional participatory futuring techniques such as public Delphis or future search conferences. The "hands-on futuring" through full-scale prototyping facilitated this engagement and may have compensated for the lead users not being the pinnacle of global lead users but local ones. For these kinds of participants, turning inspiration cards into simple category markers did not stale the participants' capacity to envision. People that display considerable lead-user characteristics hold extensive solution and trend information already, and complex creativity props could have stood in the way of elaborating what they found important. Also, keeping these participants working alone for much of the time ensured they had due opportunity to express their assessments and solutions in their deep competence area. Finally, had many participants not stressed the difficulty of attending longer than a one-day workshop, the trend exploration and solution identification could have been pursued in a more repeated fashion.

Design Workshop as means: Developing a Novel Online Portal for Maternity Wards

The maker space workshop was a very demanding and even academically new opening in which all aspects of design participation required thorough and painstaking considerations. Not all design participation is like that. The workshop we examine next exemplifies design participation in the late phases of concept design and in a workshop setting that needed designing beyond *bricolage* intermediate design, yet it could be realized without extensive additional investment or consideration to ensure the design participation will succeed. We still see all types of design participation work being involved but pursued in a considerably less painstaking manner.

Design Participation Context and Framing Work: Maternity Ward Online Portal

One of our past projects concerned the online services and information provided by the maternity care clinics of the City of Helsinki, as part of Aalto University's broader "Living + " initiative with the City of Helsinki (Hyysalo, Greger and Hatami 2012). The project responded to the two-fold changes that had taken place in the preceding years in the life of Finnish families with children. First, the internet and its discussion forums had exploded the availability and ease of finding information related to pregnancy and infant care, albeit with an alarming proportion of unreliable information. Second, experiments with self-service stations in care clinics had shown that many parents had become competent in carrying out routine measurements of their babies. This was particularly so after the first few months with infants and after their first child, and such parents preferred to just fill in the basic measure information independently and fully devote their appointments to discussion with the maternity care nurses. With better online services, they could do routine measurements from home as modern scales and measures allowed for the needed calibrations. Redesigning the online resources of maternity wards thus became framed both as a quality assurance as well as providing new (self)service opportunities.

Relevance Work

The framing work was intricately tied up with relevance work, as our brief was being developed by a team of doctors and nurses from the city, including a senior manager from the board of health division. To anchor the design project firmly on healthcare and parent realities, and to acquaint the master's students participating in the project (led by a professor and senior service designer) in health care settings, we began the project with a three-week design ethnography and interviews in two city maternity care clinics and among families with small children. The findings from the initial study phase were consolidated into an initial concept, which was discussed with the healthcare team. Upon encouragement and some helpful comments from our partners in the healthcare units, the concept was elaborated on in more detail and turned into a wireframe prototype that allowed specifying the "logic" of the online resource as well as its exact architecture. This way of working was motivated by the main challenge that had become evident in our pre-study: the content of the information resource was relatively unproblematic but very tricky to prioritize adequately and even trickier to achieve within a clear architecture for the portal with sufficient prioritization. To this end, we decided to turn to collaborative design workshops because we realized that decisions about how to arrange and prioritize the information could be much better elaborated on by the implicated people themselves rather than by us on behalf of them.

Selection Work

The codesign sessions that focused on improving the design concept and prioritizing the contents presented were divided into two—families and professionals—reflecting the quite different perspectives and knowledge asymmetries that existed between these groups. For the families' workshop, we invited several parents with infants based on the design team's acquaintance with them. For the professionals' workshop, we invited three maternity nurses and one doctor working in our contact maternity ward to get different professional perspectives on the portal and its contents. What we thus had was a relatively straightforward and convenience-based selection, as the information needs, representativeness, and expertise requirements appeared generic at this stage. Furthermore, there was the possibility to expand the participant group after these workshops if need be.

Constituency Building Work

The constituency building work remained relatively modest as well, as the healthcare or family participants were not enrolled in design or facilitation work beyond their participation in the workshop and information provision preceding it. Because design was quite new to the City of Helsinki healthcare division at the time, the project was part of the efforts to initiate design-related constituencies. It was believed that the commission from the top of the city's healthcare division would result in sufficient uptake of both insights and designs (which eventually proved not to be the case).

Intermediate Design Work

The intermediate design work was comprised of turning insights from the design ethnography and interviews into a preliminary design concept. The concept structured the information according to the week of pregnancy and the age of the child after birth according to key steps defined in the maternity health checkups. The service concept was then turned into a comprehensive set of wireframes, which were printed onto A3 paper to allow for flexible display and movement within the service architecture. The considerable amount of information content that was already identified as likely to be incorporated were printed onto smaller rectangular paper sheets that could be placed onto the wireframe sheets. In addition, four colors of Post-it notes—red for problem, green for good solution, yellow for neutral added information, and blue for priority information—were used to complement missing information content (see Fig. 2.9). These elements were placed onto a large tabletop together with pens and additional paper. A notetaker focused solely on writing a memo, and a facilitator ran the session and introduced the service wireframe and its sub-areas according to the script of the two-hour session, which was written beforehand—this setup allowed for 3–4 participants at a time to effectively participate in the designing.

Collaboration Work

The collaboration work began with an introduction of the aims and suggested flow-through of the workshop and the introduction of the overall concept, followed by the first discussion. The facilitator next guided the four participants onto the landing page and initial screens of the wireframe, keeping the session in constant dialogue rather than usability testing mode (Buur and

Fig. 2.9 Collaborative design setup in maternity ward codesign workshop. Student group Mäkinen, Abhisek, and Jung-youn, Instructors Hyysalo & Greger: Lapsen kanssa/Co-designing the future health IT project, Aalto University

Bagger 1999). She then proceeded to walk through each designated pregnancy/baby's life week with the participants to prioritize and add to the information that should be provided for that week and to order it physically onto each week's sheet. The session lasted approximately two hours and ended with a check-up walkthrough of all the sheet information and a final discussion on the concept and its merits. At this point, the whole wireframe concept had been elaborated on regarding the key information, its phrasing and location, and order of presentation. The team then took about a day to polish and integrate the information into the digital design and to resolve the few differences that ensued from the families' and professionals' codesign sessions.

Outcome Work

The outcomes of the workshops were integrated into a design concept called "with child" by the master's students. Several changes were made to the concept's architecture, and its information prioritization was further refined. The finalized concepts were then presented to the board of the City of Helsinki health services, which discussed them enthusiastically for a half an hour in their board meeting. There was a strong sense that the concept and the way it had been developed captured what the client and different user groups had wished for. However, just half a year later, the city's health and social care divisions entered a massive merger that halted any and all smaller projects that were not already in production, resulting in "with child" becoming shelved and eventually never becoming implemented.

Reflecting on the Maternity Care Workshop

The use of participatory workshops was assessed by the team to have been extremely effective, as just two 2-hour workshops and one hour for polishing resulted in verified information needed to complete the alpha version of the design concept. This could have been done through more traditional human-centered design, in which designers would have determined the informational content and forms themselves and then usability tested the result either as cognitive walkthrough screens or through test-tasks and then iterated the design based on the result. Our estimate was that six rounds of iteration and thus about three weeks of time would have been needed to reach the same point that was now reached in just two days.

At the same time, it was evident that our project had underestimated the amount of relevance work and outcome work needed for the design concepts to become implemented by the city healthcare division, which went into a merger with the social care division right after the program, resulting in the halting of all other development efforts. Had this been known and properly planned for in the framing and relevance work, the design concepts could have been developed further such that they could have been outsourced for development and had been ready to launch after the merger was complete.

Chapter Conclusions

This chapter has concretized design participation work for single workshops through two examples. What we have shown is that one should seek to arrive at a meaningfully designed event in which the participants can contribute to elaborating on and/or solving the problem at hand. One cannot and certainly should not try to "design the participation" (cf. Redström 2006)—that is, tightly designate who the participants are, what exactly they are to do, and what outcomes they should produce—for what then would be the point of having people participate in the first instance beyond a self-legitimizing tick-the-box exercise? Neither is a workshop a detached space–time or a detached design object; rather, its adequate design results from meshing the methods with design work and mundane work and strategic considerations of each other. The types of design participation work that we elaborated—framing work, relevance work, selection work, constituency building work, intermediate design work, collaboration work, and outcome work—help to illustrate what needs to be considered (at least).

To show the effects of these considerations and how they operationalize in the nitty–gritty design decisions, this chapter elaborated on the design process for two half-a-day to day-long workshops that had different contexts, underpinnings, and eventual aims. Through these examples, we gained a sense of the considerable differences there are in different design contexts even for workshops. The maternity ward workshop centered around intermediate design anchored to the wireframe prototype (an evolving object as the prime mediating means between designers and domain experts), whereas the library maker space involved a considerably more complex intermediate design for gaining the needed trend and solution information.

The design timeframe was also an important difference: designing solutions going into production within months or after several years made the maker space workshop more demanding. The work involved in participant attraction and selection was also very different (albeit still much more straightforward than in the design community formation described in Chapter 5). In both cases, the work types also intertwined. For example, particularly the relevance and outcome work in both workshops affected who would be the relevant participants, and this in turn affected the requisites for the intermediate designs and designing.

In terms of outputs, even as these workshops required considerable resources to design, run, and analyze, the insights gained would have been difficult and even more costly to attain by other means. In normative terms, directly engaging with knowledgeable users was a considerably more democratic way of engaging than usual expert surveys, interviews, or even observations. One can always ask if a still wider participation from, for example, maker communities and perhaps the inclusion of non-maker citizens could have worked just as well. Yet having participants who were deeply immersed in current and future work practices did allow us to work without further prompting and gave the participants an unfiltered way to represent their case.

An important aspect of both workshops was the material make-up of the workshop, which made it possible for the participants to directly engage with the design matters at hand, furthering the design concretely. The material arrangements in both workshops were also pivotal to the effectiveness of the workshops, allowing the participants to shape the design matters in just hours in contrast to much longer timeframes that would have been needed if only discursive participation formats would have been used (cf. Ehn and Kyng 1992; Eriksen 2009; Ertner et al. 2023). In the maker workshop case, the invested participants and progressing material embodiment next made it possible to proceed to successful implementation of the pilot maker space in which many of the participants as well as broader sets of users could participate in elaborating on library maker facilities.

We shall next move into what follows when design participation is extended temporally into the design-in-use period (Chapter 3), when its target system is expanded to societal sector-wide sociotechnical systems change (Chapter 4), and when it is expanded to purposefully reach wide and diverse sets of participants through a mix of formats (Chapter 5).

CHAPTER 3

Extending Participation into Use Time and Design-in-Use

Introduction: From Active Use to Extending Design into Use

In this chapter,[1] we move beyond the workshop and first examine what people do and alter when they engage with products, services in their work and everyday life settings. To this end, we introduce a taxonomy of active use that synthesizes the somewhat disparate research literatures on the topic. We then move onto elaborate one of the implied avenues for design participation, namely extending design and participation into use time through an approach called aging together and the implications it has regarding infrastructuring with everyday communities.

We begin by noting how the dichotomies between "designers" and "users" or "designers" and "non-designers" no longer hold (Buchanan 2001; Woodhouse and Patton 2004; Manzini 2015). People traditionally thought of as "consumers" or "users" of designed products and services actually engage in many design-related activities beyond routine purchasing and using artifacts. People today do all manner of their own designing and making of

---

Coauthored with Andrea Botero and Cindy Kohtala.

[1] Earlier versions of analyses that are reworked into this chapter have appeared in Kohtala, Cindy, Hyysalo, Sampsa and Whalen, Jack. (2019) A taxonomy of users' active design engagement in twenty-first century. *Design Studies 67*, 27–54; Botero, Andrea and Hyysalo, Sampsa. (2013) Ageing together: Steps towards evolutionary co-design in everyday practices. *CoDesign: International Journal of CoCreation in Design and the Arts* 9(1), 37–54.

S. Hyysalo, *Design Participation*,
https://doi.org/10.1007/978-3-031-95437-5_3

goods and services—tweaking, redesigning, appropriating, hacking, and "prosuming" (Knott 2013)—which means both producing and consuming (as in early webpages or today's open hardware by "makers") and even innovating. It is widely recognized that makers, prosumers, and user innovators are taking more roles in design and innovation, both as individuals and as part of broader design networks and communities (e.g., Margolin 1997; von Hippel 2005, 2016; Oudshoorn and Pinch 2003; Rohracher 2005; Sanders and Stappers 2008; Hyysalo, Jensen et al. 2016; Richterich 2020; Rossi et al. 2021). The range of these practices involving actively design-engaged use is thus not only diverse but also significant economically, societally, and environmentally. Moreover, it carries significance for the design profession as well (Campbell 2005; van Abel et al. 2011; Redström 2006, 2008; Bødker et al. 2004; Green et al. 2019).

In both design research and practice, whole repertoires of approaches have been developed over the last three decades that are premised precisely on how people using the products and services take design further—that is, **design-in-use**. In participatory design, Henderson and Kyng (1992) looked at the advantages of continuing collaborative design after implementation based on design adjustments made by users, and not only in the envisioning stage (Ehn 1992). This has subsequently developed into design-in-use approaches including co-realization (Hartswood et al. 2002; Büscher et al. 2009), meta-design (Giaccardi and Fischer 2008), aging together (Botero and Hyysalo 2013), and gradual system expansion strategies (Whalen and Dobrow 2012; Johnson 2013). These approaches have widened and fed into design that aims toward infrastructuring, the active doing of infrastructures locally and across locations, to empower individuals and communities in their social and material pursuits (e.g., Agid 2016; Büscher et al. 2009; Botero 2013; Karasti et al. 2020; Naito and Botero 2024). The rise of hacking and digital-physical making, in turn, is linked to open design initiatives and open hardware projects (van Abel et al. 2011; Tooze et al. 2014; Aitamurto et al. 2015; Bakırlıoğlu and Kohtala 2019); peer content creation and crowdsourced design and innovation strategies (von Hippel 2016); and designing with and for maker communities (Ratto and Boler 2014; Seravalli 2014; Vezzoli et al. 2021). In all these research and practice approaches, the key underlying question is: what are people capable of when engaging with designs-in-use, and how can their engagement be supported and built upon?

Designers are not alone in seeking to understand and engage with these designs-in-use. Many disciplines beyond design have contributed significantly to our understanding of how people engage with (forms of) design by observing and distinguishing what they do—their active use. That is, scholars have looked at what modes people adopt in engaging with products and technologies (e.g., Eglash 2004; Campbell 2005; Woodhouse and Patton 2004)—whether they innovate or not and how (e.g., von Hippel 1976, 2005, 2016) and how they participate in design, production, and consumption (e.g., Campbell 2005; Shove et al. 2007; Simonsen and Robertson 2013). This understanding can be conceptualized for clarity as "active use," as it highlights what is enacted *while using*, rather than in terms of design-in-use, which also includes the practices in which designers do an important part of their designing during use time. In what follows, we will use both pairs of concepts, active use and design-in-use, to elaborate on ways in which both active use and design-in-use help us understand more nuanced ways to extend participation.

Empirical findings from the various disciplines pointed to the above have gradually been worked into typologies that clarify design-use relations in a particular context and from a particular research position. Sanders (2006), for example, demarcates between "customer, consumer, user, participant, adapter, maker, co-creator"; Eglash (2004) between "consumption, reinterpretation, adaptation, reinvention, and production"; de Jong et al. (2015) between "routine use, user modifications, and user innovations"; while Campbell (2005) differentiates between "cultural dupe, personalization, customization, and craft consumption." Such demarcations typically serve well to advance specific arguments, such as what demarcates craft consumption from mere personalization (Campbell 2005).

At times, these typologies feature partially overlapping terminologies for people's design engagements during use time and thus result in some unhelpful redundancy of terms. Many typologies also partially diverge, covering different areas in active engagement with designs during the use time. Complementarities and disparities between the proposed ways of classifying active use and design-in-use have resulted in the crossbreeding of simple typologies into more comprehensive frameworks (e.g., Botero et al. 2010; Botero 2013; Juntunen 2014). We therefore see that there is potential for important insights between fields related to design participation.

If taken together at face value, the varying existing typologies produce a conceptually blurry depiction of active design engagement, which hampers both researchers and practitioners in their engagements with active use. To clarify rather than add to the muddle, we next elaborate an analytically ordered taxonomy of active-use-related phenomena to duly clarify the breadth and depth of active design engagement during use time, which, in turn, supports an expanded articulation of design participation.

Typologies of Active Use and Design Engagement

In elaborating on this proposal, we limit our focus to what people do with products and services when they engage "first-hand" with their material qualities—by themselves or with their peers—as non-professional designers or producers. This means we are excluding ways in which people affect designs as informants or aids to professional designers. We also do not include engagements related to filing complaints, lawsuits, boycotts, demonstrations, or other indirect mechanisms to spur designers or producers into action (e.g., Olsson 2004; Bovaird 2007; Pollock et al. 2007; Smith 2012). Nor do we include general theories of sociomateriality or post-human sociology (e.g., Latour

2005; Dant 2005; Engeström 2000), as these theories do not address in detail first-hand engagement with the material qualities of products or services.

To get started, let us first recount a set of well-cited, simple one-axis typologies that address the continuum from passive use to what authors considered the most active engagement with design-in-use. This helps to see some clear commonalities as well as different dimensions of what people are "active with" regarding design. Once these basic parameters are in place, we can cross-compare and provide a first synthesis of the typologies.

In design studies, the focus has been to understand the differences between how professional designers and lay people engage in design (Sanders 2006; Taffe 2015) and how professionals can capacitate lay people to design more widely and deeply (e.g., Bødker et al. 2004; Simonsen and Robertson 2013). Following Henderson and Kyng's (1992) seminal work on the advantages of continuing collaborative design after implementation, suites of design-in-use approaches have emerged. Yet probably the best-known typology in design research is Sanders's (2006) model of user involvement that has been adapted and expanded by others sharing the interest to identify opportunities for professional design practitioners to codesign with users (as "lay designers") in developing commercial products. Hermans (2015) has adapted Sanders's framework further, as illustrated in Fig. 3.1.

Research on user innovation since the 1970s has focused on whether and how users have innovated or not, differentiating whether users have ***modified*** existing products or ***innovated*** entirely new material solutions or techniques (de Jong et al. 2015; Hienerth et al. 2014; Halbinger 2018). Similarly, information systems and human–computer interaction (HCI) research has since the 1980s paid attention to user engagement with design, exploring active use in IT appropriation (DeSanctis and Poole 1994; Orlikowski 2000; Alter 2006). DeSanctis and Poole (1994) presented a thorough HCI framework on the most common "appropriation moves" with organizational software. These moves range from ***appropriating as is***, ***substituting*** software use by other means, ***combining*** different software to achieve aims, ***enlarging*** the functionalities to new uses, ***constraining*** the use only to some aspects of the system, and ***contrasting*** software use with other means. Further moves, such as repurposing appropriations and creative uses and their situational and positional

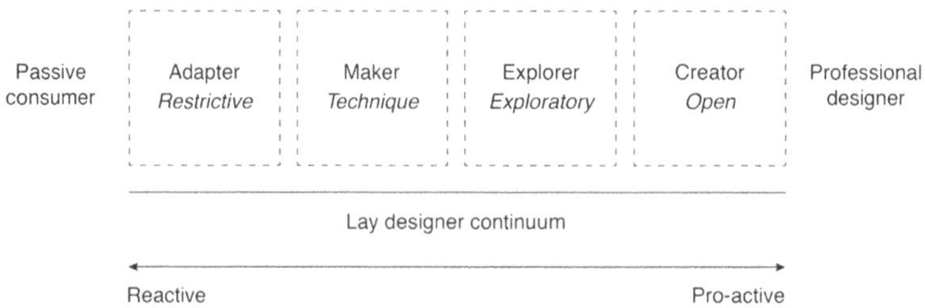

Fig. 3.1 "Lay designer continuum" (Hermans 2015, p. 157)

underpinnings, have been further elaborated over the years (e.g., Salovaara 2012; Liikkanen and Salovaara 2015).

Just as importantly, in consumption studies, there are several categorizations related to active use. Campbell (2005) differentiates between "cultural dupe, personalization, customization, craft consumption," which notes how people both alter objects and alter *uses* and *meanings*. The domestication framework (Silverstone et al. 1992; Berker et al. 2006; Hartmann, 2023) draws out different aspects of active use and engagement with design through "appropriation, objectification, incorporation, conversion." Here, appropriation (understood narrowly) takes place when a person takes the artifact into their own *economic* possession, turning it from the object being in a "commodity state" tradable in the market and into a non-commodity state (Kopytoff 1986). Objectification is an activity whereby the good is integrated into the *physical context* and its existing orderings; incorporation then marks how the functions of a good become intertwined into the users' functional purposes and related *functional meanings*. Lastly, conversion signals how the previously "alien" good is converted to convey meanings about its owner in *interpersonal meanings* through expressions of, for example, wealth, style, or dispositions.

Finally, science and technology studies (S&TS) feature typologies regarding how individuals engage with designs. Eglash's (2004) typology starts from reinterpretation, i.e., change in *meaning*, and continues to changing *use* and to changing *materiality*. In this typology, Eglash emphasizes social power, marginalization, the black boxing of technologies, and citizens' strategies to change these power relations.

Taking stock of these distinctions, we can see that these typologies explicitly move from what their authors regard as conventional "consumption" toward "prosumption" or innovation. While there is some variation, these models suggest there is a set of distinctions regarding the *intensity* of active use. Compiling these various intensities would result in 6–8 overlapping categories; we thus propose these can be analytically redacted to three distinct intensities in active use where the differences noted in the literature are important (Fig. 3.2).

If "use as is" is routinized use of a product or service by a user, or use as implied in marketing, manuals, and the like, then on our "axis," the next important intensity of engagement with design is "active and mildly adaptive use that involves tweaks of some kind," which we call "active use." The next intensity of engagement with design is "locally innovative designs and

Fig. 3.2 The degrees of intensity in active-use phenomena

modifications," which we label as "user design." Finally, we demarcate "new-to-the-world design" as the most intense engagement with design, which we label "user innovation" (Fig. 3.2).

The cross-comparison of the typologies in the literature reveals different aspects to which this "degree" of intensity of active engagement may relate. We have been talking about how people can change how they *use* designs such as making adjustments or workarounds (Alter 2006), technique innovation (Hienerth et al. 2014), or "exaptation," where a feature or product takes on a new function not originally intended (Andriani and Cattani 2016) (see the top vertical row in Fig. 3.3). People also design, redesign, alter, and innovative the very *objects* themselves (second row in Fig. 3.3). We also noted that people change the *meanings* connected to designs (Fig. 3.3, third row), as in new semantic associations (Eglash 2004; Pfaffenberger 1992). Consumption studies, design studies, and S&TS draw attention to user-made alterations to *settings* or in digital cases *local platforms* where a certain use happens to be in this home or that digital site (Fig. 3.3, fourth vertical row). To clarify, from the perspective of actively engaging with design regarding uses, object, or meaning, the local setting is the setting in the context. But when the use of a novel design leads to alterations in this context setting to accommodate the design, the context itself becomes an object of design action, foremost pursued when users do infrastructuring (Botero and Saad-Sulonen 2022). Such alterations can be found in intertwining novel designs in existing homes, as documented in domestication studies (Silverstone et al. 1992; Berker et al. 2005), in extensive repair and do-it-yourself practices at home where not only novel designs but surrounding spatial arrangements are altered (Shove et al. 2007), or bricolage, when assembling and remixing elements by hand, in digital and physical settings (Büscher et al. 2001; Usenyuk et al. 2022). Hence, to dispel a fair amount of confusion over different typologies, Fig. 3.3 presents a minimal taxonomic framework for discussing active design engagement.

Let us next focus on *collective* forms of users' active design engagement that have long been documented (e.g., Allen 1983; Schiavone and Esposito De Falco 2016) and whose prevalence, forms, extent, impact, and visibility have been boosted by digital connectivity and availability of easy-to-use and shared digital design tools (von Hippel 2005; van Abel et al. 2011). For instance, Botero et al. (2010) have outlined forms of active user engagement in digital community design and associated supporting design processes that can be carried out either by peers or by professional designers or other service providers. The authors split their framework into increasingly intensive social design forms (creating workarounds, making social agreements, and fostering evolution in social practices) and increasingly intensive technical user engagements (integrating, personalizing, aggregating/remixing, assembling components, using modules and libraries to design, and programming new libraries).

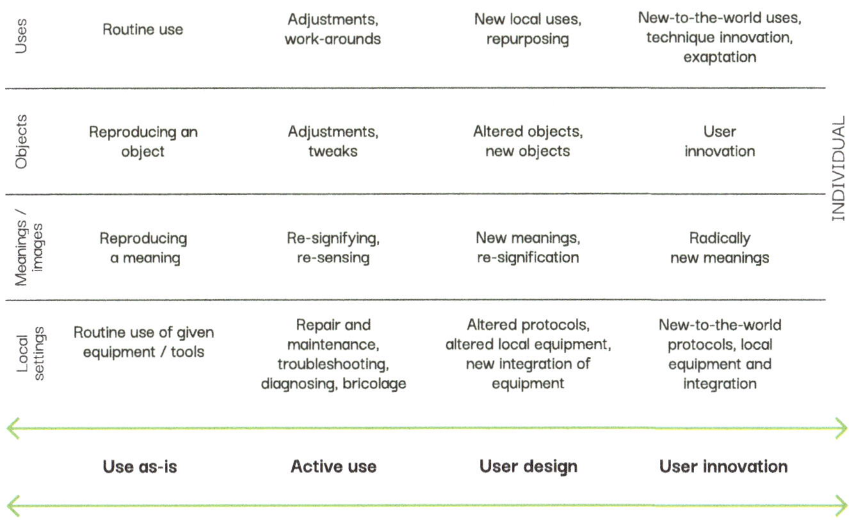

Fig. 3.3 A minimal framework for discussing active design engagement in a given setting

There have been similar findings regarding users' and makers' active engagement in design, extending from tools and construction to maintaining communities and organizations, their routines, social practices, rules, and procedures. These engagement patterns have become especially salient in various open design, open source, and "DIY maker" peer social groups (Kuznetsov and Paulos 2010; Toombs et al. 2014; Kohtala 2018). These groups often assemble in devoted open-access spaces (e.g., Fab Labs, maker spaces, hackerspaces) to use tools and equipment to design and make their own artifacts, mobilizing community governance models that imitate modes established in Free/Libre/Open Source Software (FLOSS) development (van Abel et al. 2011; Marttila et al. 2014; Aitamurto et al. 2015). In these open design initiatives, participants are involved in designing and sharing designs (Tooze et al. 2014; Özkil 2017), as well as in altering the design of physical settings for their actions (Kohtala 2017, 2018; Naito and Botero 2024).

Participants adopt a range of strategies to engage themselves in technology and product development, coproduce services, and involve others in community work. Some of these engagements are routine, others introduce alterations, and still others introduce new-to-the-world ways of acting. Building and maintaining *communities and organizations* become requisites for collective engagement with design (van Abel et al. 2011; Aitamurto et al. 2015; Kohtala 2017; Bakırlıoğlu and Kohtala 2019). As in more traditional settings, peers mediate each other's design engagement through brokering contacts, facilitating learning, and configuring systems (Stewart and Hyysalo 2008; Hakkarainen and Hyysalo 2016), as well as through acting as volunteers and community organizers (Johnson 2013). Some act as "warm experts" from whom ignorant and embarrassing questions can be asked (Bakardjieva 2005), as "configurers" (Okamura et al. 1994), or more widely as "local experts" to whom members in a local community can turn for problem-solving help beyond the capabilities of themselves and their closest circles of peers (Stewart 2003). Such peer-to-peer and peer-to-community actions are crucial for participants' control over the environment in which their design engagement takes place (Arnstein 1969; Cardullo and Kitchin 2019).

A differentiation is also needed between the local settings and the wide, sometimes literally global, *platforms* that connect a wide range of local settings within a given domain (Benkler 2006; Botero et al. 2010; Özkil 2017; Kohtala 2016). Altering and setting up trans-local settings take place in physical goods and more restrictedly available technologies such as proprietary digital services; however, it then makes more sense to talk about *interaction arenas* than platforms (Hyysalo 2010, 2016; Johnson 2013; Hyysalo and Usenyuk 2015).

The technology-mediated strategies of social control and the counter strategies offer insights into yet another set of phenomena. Pfaffenberger (1992) offers a typology that spans "regularization—counter significations—counter appropriations—counter delegation (non-use, modifications, hacking, reuse)—reconstitution." In this schema, reconstitution means actively reshaping technological production processes or artifacts guided by a conscious ideology such as that seen in much of the open and free software movement, the open design, collective production of "counter-artifacts" such as Linux or Mozilla Firefox that are free of the dominant industrial regime and its regularization strategies (Corbett 2014; Kohtala 2017). These ideologies are co-created and are similar to "imaginaries," a group's or society's collective and partially materialized aspirations for future science and technology (Hyysalo 2006; Flichy 2007a, 2008; Jasanoff and Kim 2015). Successful reconstitution of an imaginary or ideology involves the alteration, design, and innovation of both objects and symbols, not only in creating counter-artifacts but also "counter-contexts" (Pfaffenberger 1992). Reconstitution strategies can be readily observed in how alternative forms of production come to be taken as the only legitimate one in the organic food production movement (Durrant 2014); community energy initiatives (Smith et al. 2014; Nielsen 2016; Hyysalo 2021); and subculture communities such as alternative hackerspaces and avowedly "green" maker spaces (Jeppesen et al. 2014; Toupin 2014; Smith 2017). The *ideology* or *imaginary* motivates these collective endeavors but are also reworked and altered by active users during the process (Gregory 2000; Flichy 2007a; Jasanoff and Kim 2015; Stein 2017; Kohtala 2017).

These considerations imply interpersonal and trans-local dimensions that provide a further dimension to active use and design engagement, which we add to the taxonomy in Fig. 3.4 as communities and organizations, imaginaries and ideologies, and global platforms or other between-setting interaction arenas.

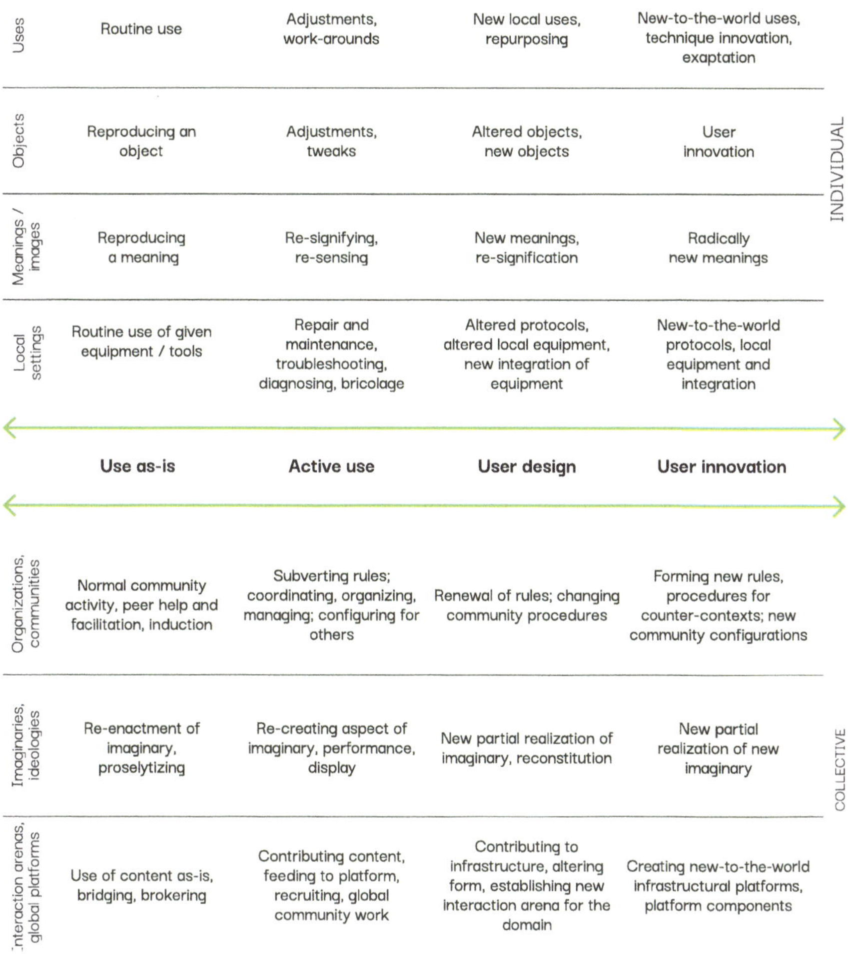

Fig. 3.4 Varieties of active design engagement

Active-Use Taxonomy at Play: Peer-to-Peer Open Design Communities

Let us elaborate on the taxonomy of active use through real-world examples from a particular domain that of sustainable maker spaces and open design initiatives, based on a four-year ethnographic study by Kohtala (2016, 2017). As introduced in Chapter 2, open design groups, commonly called makers, often collaborate in shared community workshops called Fab Labs or maker spaces equipped with small-scale, digitally controlled production equipment such as milling machines and 3D printers. They are thus prosumers, both producers and consumers: they engage in object design (designing and making physical artifacts), as well as community design (designing events, interactions, and community governance models). In such horizontal peer-to-peer communities, the boundaries between "designer," "user," and "organizer" are fluid and shifting. In addition, Fab Lab settings bridge both physical, tangible materials and digital artifacts and infrastructure. These settings feature all dimensions and intensities of active-use phenomena, condensed in Fig. 3.5.

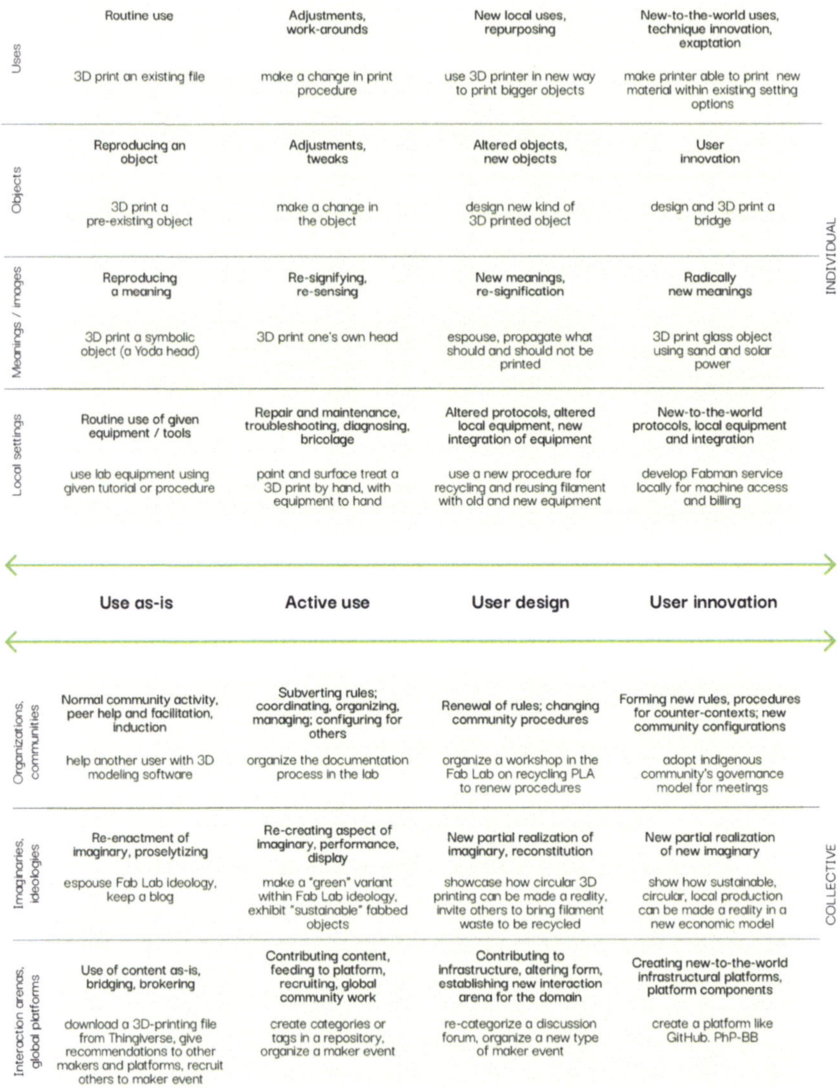

Fig. 3.5 Varieties of active design engagement in peer-to-peer open design initiatives

Let us start with the USES row and the example of using a 3D-printer in a Fab Lab. Makers can use a printer "as is," as routine use, to print out an existing design file. When they make a few tweaks and adjustments such as experimenting with settings and changing the existing print procedure, this denotes "active use." Some makers use printers in a new way, setting them up on hydraulic lifts, for instance, to print bigger objects that require design work beyond what is done in a moment constituting "user design." Some makers engage in technique innovation and new-to-the-world uses, using the printer to print in another material such as porcelain, comprizing "user innovation". Innovation by exaptation, using objects in entirely new uses they are not designed for, is also common. For instance, makers may change the purpose of video game devices such as the Xbox Kinect from motion sensing for gameplay to 3D scanners in order to create 3D models.

Let us move to the OBJECTS row. Makers often reproduce objects as is, such as 3D-printing an artifact from a pre-existing design file, which is "use as is." When they actively tweak the object, making a change in the object to personalize it in some way, this is "active use." Many makers start from scratch and design a new kind of 3D-printed object: "a user design." Some makers innovate by creating new-to-the-world objects, such as a full-size 3D-printed steel bridge, comprising a "user innovation."

Makers associate MEANINGS AND IMAGES with actions and interactions, which serve to legitimize them, and this is an important element in design engagement beyond objects and uses, as we have learned from consumption studies. Many first-time visitors to a Fab Lab simply 3D print an object; a favorite one to print is the Yoda character from Star Wars, which acts as a "geek" symbol in maker culture. Someone 3D-printing in the Fab Lab in this way is reproducing a meaning: "use as is." Some makers then seek to explore the meaning of 3D-printing further for themselves, re-signifying by 3D-printing a scan of their own head: "active use." Some makers seek to more significantly redefine the meaning and purpose of 3D-printing, due to their own concerns with, for example, plastic waste: they propagate new meanings of ecological 3D-printing through eco-material experiments or recycling plastic waste into new filament: "user design" in meanings. Makers have also created radically new meanings concerning the role of 3D-printing for themselves (and possibly the wider community as well), such as producing glass objects using sand and the sun with a solar-powered 3D-printer (Kayser 2011): a "user innovation" in meaning of engagement.

Turning to a Fab Lab as a LOCAL SETTING, people using the lab engage in routine use of the given tools and equipment and use a given tutorial or procedure: "use as is." Some makers use the given tools and materials but may remix materials and components or undertake repair and maintenance, troubleshooting or diagnosing tasks. Repairing a piece of equipment or painting or surface-treating a 3D print by hand, with equipment at hand, for instance, suggests bricolage in the setting, one form of "active use" related to setting. Makers also alter protocols or equipment more significantly. For example, they use new procedures for recycling and reusing 3D-printing filament, with old and new equipment including office paper shredders, forming "user design." And makers develop new-to-the-world protocols, local equipment, and integration, such as locally developed digital applications for machine access and automizing machine time billing such as the "Fabman" service, first developed as a maker space management system in Happylab, Vienna, and now a commercialized software suite used by Fab Labs globally. This is an example of "user innovation" in a local setting.

Now we turn to the collective components of the taxonomy (Fig. 3.5), where we continue to illustrate intensities of active design engagement through the example of sustainable maker spaces and open design initiatives. These initiatives are not commercial manufacturing services: people join them in order to join a community and engage in design and designing that is socially oriented.

In the ORGANIZATIONS AND COMMUNITIES row, makers who routinely engage in a Fab Lab engage in normal community practice by adhering to the lab's rules and routines and identifying with the organizational culture. They engage in peer help, facilitation, and induction by, for instance, helping others with 3D-modeling software as a routine community activity. This would class in the "use as is" category. Makers may then break or bend a rule in the lab, alter a procedure, or initiate organizing or coordinating for others. Some makers, for instance, take on a task of (re)organizing the documentation process in the lab, which is "active use." Local designing in and for the community also occurs, renewing the procedures or rules by, for instance, creating their own house rules for the lab. A group who wants to change procedures to orient the community to more environmentally conscious practices might start by organizing regular workshops on recycling PLA filament (polylactic acid, a bioplastic) instead of sending it to a landfill. This would be "user design" regarding organizing the community practices. Makers also introduce new-to-the-world elements into the community or into the practices that constitute it, such as adapting local governance models (e.g., meeting protocols) to ones found with local indigenous communities' models or community economy principles (Neale and Hobern 2017; Smith 2020), comprising "user innovation." In the example of recycling PLA, user innovations for the community have entailed recruiting others into open innovation practices with free sharing of equipment designs and instructions, steering material flows toward circularity, and offering space for experimentation and new revenue streams as entirely new community configurations (e.g., Spekkink et al. 2022).

We then move to the IMAGINARIES AND IDEOLOGIES row. As mentioned earlier, some communities such as open design initiatives operate like social movements. They convey ideologies to attract and mobilize participants, and they co-create, use, and adapt visions of desired futures that are publicly performed to align strategic partnerships and gather resources (Kohtala 2017; Stein 2017). Makers simply re-enact an imaginary when they advocate and proselytize, for example, writing a blog that reproduces Fab Lab ideological content to others. In the taxonomy, this is "use as is." Making a variant on the ideology, such as emphasizing an aspect of the ideology they value related to ecological issues or exhibiting "sustainable" fabbed objects in the lab through performance or display, is "active use." Some makers start to make this new aspect of the imaginary more "real" by doing and making, conveying the new aspects through symbols and objects, and inviting others in, such as by showing how circular 3D-printing can be made a reality by showcasing the lifecycle of 3D-printing bio-based filament to the community and inviting them to bring 3D-printing filament waste to the lab to be recycled. This would be "user design" as a new partial material realization of sustainable making imaginary. Even further, some Fab Lab collectives start to create and (partially) realize a completely new imaginary for peer-to-peer open design, aiming to show how, for example, sustainable, circular, local production can be made a reality in a new economic model involving their own local currency. This engaging with a new imaginary also involves collective experiments in the lab and discursive items such as manifestos, displays, or texts, thus a further intensity matching "user innovation."

In the final row of the taxonomy, we examine INTERACTION ARENAS AND GLOBAL PLATFORMS. Fab Lab communities generally do not operate only locally. They rely on global networks of connections to infrastructures and other Fab Labs. In this category, makers participate in interaction arenas such as maker events, and they download open-source design files from online file repositories such as Thingiverse. Such straightforward action and use of content in global platforms and arenas is "use as is." Makers actively contribute to shaping the platforms, too, by for example, creating new categories or tags in a repository or recruiting and doing community work in organizing a cross-setting maker event—forming "active use." Makers also alter the form of platforms more thoroughly, and they establish new interaction arenas for the domain. Fab Lab participants have, for instance, altered and re-categorized the online discussion forums for the global community, and they have organized new types of face-to-face, cross-setting maker events that would be "user design" regarding interaction arenas. Finally, makers have created new-to-the-world infrastructural platforms or platform components such as GitHub (an open version-control development platform for collaboration) and PhP-BB (free and open-source forum software), which are "user innovations" in interaction arenas and global platforms.

It is worth noting that many active-use phenomena tend to intertwine and feed onto each other (Saad-Sulonen and Botero 2022). As an example in maker spaces in Fig. 3.6, we see two maker peers troubleshoot the design of a 3D-printer model called MOWI[2] (active use in uses), which is a novel configuration of a printer adapting others' open-source designs and is meant to be foldable and transportable on a bicycle for use in repair cafés (user design in

[2] https://reprap.com/wiki/Mowi_printer.

Fig. 3.6 Active design engagement in a sustainable Fab Lab intertwining several types of active use (Kohtala 2016)

objects). The printer was presented in a workshop in a new interaction arena, a sustainable maker event in the Netherlands in 2016 called Koppelting (user design in an interaction arena).

Active Use Beyond Open Design Communities

Peer-to-peer open design activities present a rich terrain for users' active engagement with design. This richness may be extraordinary given that open design activities feature shared common design work; alternative new production networks, technologies, physical settings, and platforms of global reach; and various ideological currents that animate users' endeavors. Such width and intensity of active-use phenomena may not be present in many other settings, yet some of these facets do feature in various digital and physical technologies as well as proprietary industrial technologies, in public services, and in more restricted peer-to-peer initiatives. To elaborate on these differences, as well as the prevalence of active-use phenomena, we can draw from the considerable case study archive on design and use of new technology, to which our 20-year research program has provided us access to either as original researchers or main thesis advisors over the past 20 years. Most of the case studies do not provide us detailed and reliable enough coverage, but there are 12 cases spread across digital and physical and academic, peer-to-peer, private company, and public-sector settings in which this can be discerned.

Figure 3.7 lists the cases, and Fig. 3.8 examines them regarding the categories of users' active design engagement and groups them in terms of digital and physical and who is the owner and producer of the technology: private company, public organization, or citizen collective. "Deep data" indicate the characteristics of the data available for analysis, whether detailed and encompassing (dot) or not (no marking), with the main difference that in the three cases with less deep data we cannot outrule that there could have been active use we just were not aware of.

Figure 3.8 indicates what forms of user engagements were observed (colored circles), not apparent according to the data gathered (white circles), and according to case. The peer-to-peer cases are labeled as P (including academic studies), private-sector cases with T, and public-sector cases with M. The setting of each case as purely physical, purely digital, or hybrid is also denoted according to the legend provided.

The figures surface a few clear insights. First, "routine use" and "active use" regarding uses, meanings, local settings, communities, and ideologies feature in all but one instance in one case (p6 in ideologies) in which the respective category was simply not relevant to the case context, and only in two cases

LEGEND	Sector	Symbol	Case name	Deep data	Case descriptor and references
	Peer-to-peer	P1	Open design makers	●	Peer-to-peer open design initiatives in Fab Labs and maker spaces (Kohtala 2016, 2017)
DIGITAL SETTING		P2	Remote terrain vehicles	●	Use and design of all-terrain vehicles, karakats, in and for remote locations (Hyysalo and Usenyuk, 2015; Usenyuk et al. 2016)
		P3	Community bicycle workshop	●	Peer-to-peer anarchist community bicycle workshop (Wu 2017)
PHYSICAL SETTING		P4	Small-scale renewables	●	Small-scale renewable energy technology (heat pumps, solar PV, solar heat, pellet burners) communities in Finland (Hyysalo et al. 2013a, b, 2017; Freeman 2015)
DIGITAL PHYSICAL SETTING	Academic codesign for peer- to-peer	P5	Digital solutions for the elderly		Co-realization of media solutions for active elderly residents (Botero 2013; Botero and Hyysalo 2013)
		P6	Exploratory design initiatives		Peer engagement in designed community interventions on "plant hotels" (Wu 2017)
	Private sector	T1	Healthcare packaged software	●	Software package for diabetes professionals and patients (Hyysalo 2010; Hyysalo and Lehenkari 2003)
		T2	Teen social media	●	Digital service development of a large teenage virtual world (Johnson 2013; Johnson et al. 2010)
		T3	Packaged enterprise software	●	Large enterprise software package for higher education sector (Pollock and Williams 2008; Pollock and Hyysalo 2014; Johnson et al. 2014)
		T4	Safety floor system	●	Healthcare safety system for the elderly developed in a living lab (Hakkarainen 2017; Hyysalo and Hakkarainen 2014)
		T5	Safety monitoring system	●	Healthcare wrist monitor safety system for the elderly (Hyysalo 2010)
	Public sector	M1	Online platform for teachers and learners		Online platform for finding and utilizing high-quality educational and audio-visual content (Hannukainen et al. 2017)

Fig. 3.7 Active-use comparison cases, short descriptors, and references

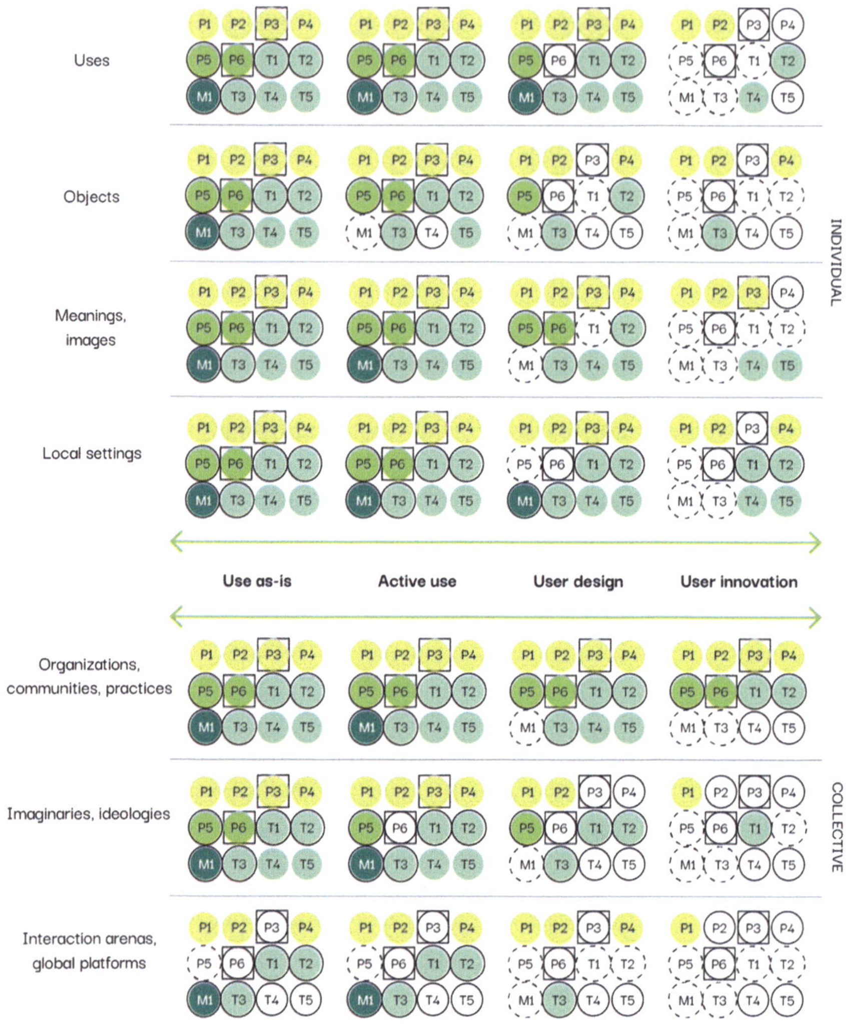

Fig. 3.8 Mapping forms of users' active design engagement across cases (colored circles mark occurrence in the case, as identified in Fig. 3.7; white circles mark non-occurrence)

there was no evidence of active use of objects. In both these cases, users did not have access to change digital objects. In addition, "user design" features widely, observed in over 10 of the 12 cases in uses, meanings, local settings, and communities and in half the cases in objects and ideologies and in half the cases (3/6) where global platforms were present. In turn, "user innovation" is rarer, featuring most frequently in organizations and practices 8/12 and local settings 5/12 but only 4/12 in meanings and 3/12 in uses and objects. Global platforms and interaction arenas feature less active user design and user innovation than other facets. Finally, and perhaps surprisingly, there are no clear differences between user engagement in cases that feature physical and digital parts apart from those that are digital only. Peer-to-peer open design initiatives, as free and open software projects more generally, appear to be unique in featuring all aspects and intensities of users' active design engagement, but overall the extent of active use phenomena across the cases is very wide.

The Import of Active Use for Extended Design-in-Use

For design and professional designers, this encompassing view on active design engagement during use time should make it clear that, in most contexts, design is not only about the intended use, but rather, it is about *what users make out of it in their real-life settings and practices*. This calls for considerations on how tightly or loosely designed in and scripted are different aspects of design. Just as well, it calls for considerations as to what kinds of access and ease are built into altering a design, as some people are likely to want to appropriate designs actively. The extent of active-use phenomena further underscores the importance of design-in-use strategies in design and development. When users pool their competences and resources, they can achieve results equal or even greater to professional designers (cf. Baldwin and von Hippel 2011; von Hippel 2016; Hyysalo and Usenyuk 2015; Halbinger 2018). This also undermines the idea that proactive engagement or alteration and innovation of meaning are distinctive skills of professional designers because the scope or quality of achievable design engagement does not seem to set maker and prosumer *collectives* inferior to professional designers.[3] But, it is important to acknowledge that active-use phenomena are concentrated in technically savvy and design-disposed people such as hackers, geeks, and user innovator collectives. Many of these groups are quite happy taking over work that professional designers used to do. At the same time, many other people are not, and will not, be quite so design savvy and will continue to benefit from designers' participatory engagements even if they engage in some or a considerable range and intensity of active use (see Fig. 3.8). And it is often

[3] In a sense, our findings recapitulate S&TS's long insistence on technologies being comprised of more than just products, services, or things. Technology use happens to be intertwined in practices and settings that are rich with artifacts and infrastructures, tied into communities and organizations, and animated by ideologies and imaginaries (Pfaffenberger 1992; Eglash 2004; Woodhouse and Patton 2004; Flichy 2007b; Ratto and Boler 2014). Even innovation by users can concern just as importantly local settings, meanings, and organizational innovation. Even though these are much harder to measure, they constitute core elements that make free innovation thrive and that are necessary for its freedom (cf. von Hippel 2016; Hartman and Hartman 2022).

these publics who suffer the most from technology that is ill-fitting to their practices as they have more limited ability to work around it.

Active use thus does not imply that design expertise or collaborative design become irrelevant, it just locates them differently in conjunction to evolution of designs in everyday life, pointing to design-in-use engagements. Let us examine one such methodology premised on design-in-use more carefully—what we call the "aging together" because of its explicit focus on engaging with and for ordinary people in their everyday life practices over long time spans. This is because most precursors of aging together lie in approaches to design-in-use in arguably less ordinary settings. Before moving to aging together, we will briefly recount two of these approaches as they articulate well some of the starting points. The *co-realization* approach was developed as a principled synthesis of ethnomethodology and participatory design (Hartswood et al. 2002) to address what Dourish and Button (1998) called the "paradox of ethnomethodologically informed design." This means the unavoidable reality of coming to terms with the fact that the implications of a new design for users' practices do not become evident by simply studying current use, as its implications for design will only be fully graspable during the new design's subsequent use (as we have demonstrated amply with the taxonomy of active use). Co-realization thus explores a more radical and shared practice between users and IT professionals, grounded in the lived experience of users in situ, and beyond the deployment of prototypes. In co-realization arrangements, designers continue to be present at the place where use unfolds—typically a workplace—for extended periods. This allows both the workers and designers to jointly realize how the system and work practice could be developed and then iteratively materialize these development directions (Hartswood et al. 2002; Voss et al. 2009). Co-realization has been pursued mostly in high dependability settings such as those in hospitals and manufacturing plants (Voss et al. 2009).

Meta-design is another alternative systems design practice that shifts design and participation toward evolving work-life contexts (Fisher and Scarff 2002; Fischer and Giaccardi 2006). A central tenet of the approach is to develop under-designed yet complete systems during "design time." These under-designed proposals are then made available to owners of problems in concrete domains. A meta-design approach includes developing sufficient flexible functionality to allow users to make redesigns during "use time" without or with minimum developers' involvement.

By taking seriously evolutionary timeframes and relationships, the above codesign approaches have extended design into use time. While many of their premises hold beyond the workplace and beyond quite design-able people, some are also in need of adjustment regarding more commonplace settings, in which ideas for "designing for practices" (e.g., Shove et al. 2007; Björgvinsson 2008; Jalas et al. 2017) give new lenses to look beyond the workplace (Wenger 1998; Shove 2007).

To clarify the difference between these design-in-use approaches and most codesign and participatory design, let us recount the two intuitive squiggle diagrams about design-use relation that we encountered in the introduction (Fig. 3.9). The upper image is an adaptation of Sanders and Stappers's (2008) squiggle model for codesign (itself an adaptation of Damien Newton's a squiggle diagram for design). The figure underscores how codesign activities take place not only during concept design but spread throughout the design cycle, as the squiggle line between designers and users (up and down) indicates. Yet in Sanders and Stappers's (2008) squiggle, the codesign activities end when the product starts to be used. In contrast, in the lower image, which appropriates the same squiggle notation, the primordial "launch" upon which design is traditionally thought to seize is just an interim point. An extended design-in-use collaboration has been prepared for in the form (or architecture) of a product or service in question as well as in the processual organization of design activities so that they can continue during the use time in users' own settings.

Aging Together Design-in-Use Methodology

The core idea of aging together is to gradually discover and make visible the design space available for a community of practice (Botero 2013). This joint discovery is being done through helping the community to grasp better the potential for (re)design they hold and to arrive at well-suited technologies-in-practice. Aging together can be approximated through a set of guideposts (G), which we introduce before illustrating them with examples from a long-term extended codesign project.

1. *(G1) Start with social practices.* Instead of assuming that design activities must occur in the studio or in exploratory workshops, aging together departs from exploring the practices of the community designed for. This premise is a starting point for most of the "design-for-practices" approaches, and its early roots can be traced to Scandinavian participatory design in the 1970s (for a review, see Törpel et al. 2009; Voss et al. 2009b).
2. *(G2) Understand the constituency, build, and explore new ones if needed.* Instead of starting from technical solutions, aging together assesses who are and who might become involved in the sought practice changes and what resources they could bring into the proverbial "table." This means identifying the design pre-setups of the involved actors, considering existing conditions and if the agendas of each party can be made mutually beneficial. This is another time-honed participatory design principle (e.g., Bjerknes et al. 1987), underscored as competency building work in Chapter 2.

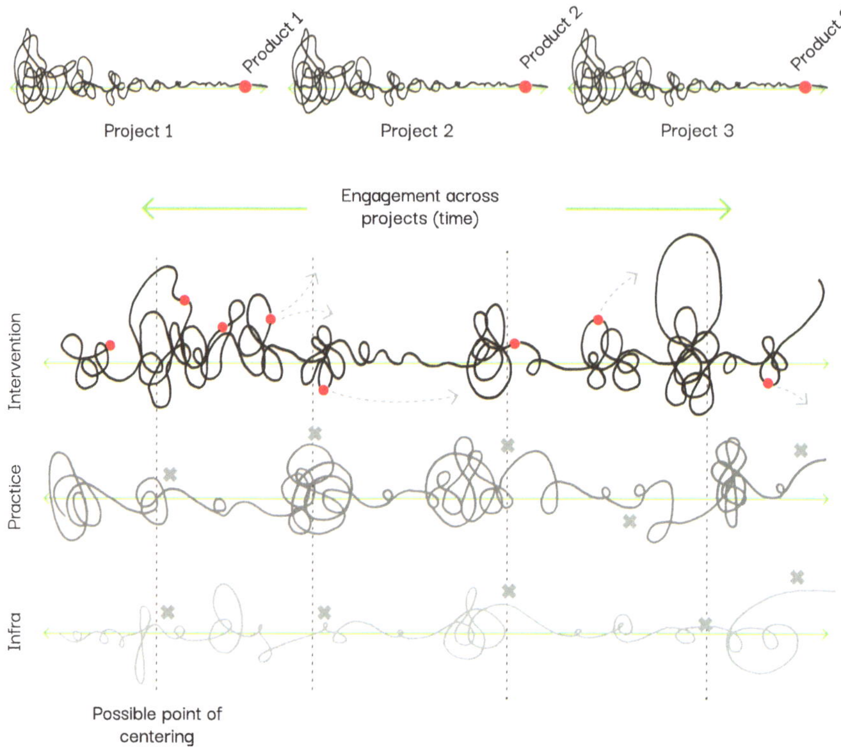

Fig. 3.9 Codesign as movement between design and use during pre-launch design, repeated for new versions and launches (upper image) and in extended design-in-use expanding to infrastructuring with the participants (lower image) Adapted from (Botero 2013)

3. *(G3) Start with access design that is small and relevant*. If the aim is for a long-term design engagement, it makes sense to gage how the collaboration works out and feels before committing deeply.
4. *(G4) Manage expectations*. While expectation management is a key part of framing and relevance work and well recognized across design with and for communities (e.g., Schuler and Namioka 1993; Simonsen and Robertson 2013), the need to address it is accentuated with collaborations that are sustained for long periods.
5. *(G5) Develop an open-ended agenda*. Aging together does not engage in design-in-use to achieve a perpetual beta killer-application but to foster contributions and competence building to improve the practices and technical and media infrastructure in a concrete community (cf. Hyubrechts et al. 2025; Karasti et al. 2020).

6. *(G6) Build scaffolds.* It is important to somehow cultivate the sense of what is possible to achieve through design. Imagination can be fostered by introducing potentially relevant concepts, materials, solutions, and good practices from elsewhere (Bucher et al. 2009).
7. *(G7) Go there, be there.* Sustained presence at user sites has been stressed through ethnographic tradition to systems design and perhaps taken to its extreme regarding design presence by the co-realization approach (Hartswood et al. 2002; Voss et al. 2009). Should the community be dispersed at the beginning, then workshops or other events in which to meet people regularly can form a good starting point.
8. (G8) Provide seed prototypes as a key part of intermediate design work for rehearsing collective and cumulative "rapid prototyping." Once a prototype system can be implemented, its design can incorporate more potential than is initially taken into use to give "double stimulus" (Vygotsky 1979) for the community members to make use of the possibilities in taking their design explorations further. Designers, in turn, should be attentive to how the community uses designs and what new shortcomings, tensions, and ideas for improvement become visible. For seeding design processes with prototypes, see, for example, Ehn and Kyng (1991), Fischer and Ostwald (2002), Voss et al. (2009), Hagen and Robertson (2010), Björgvinsson et al. (2012).
9. *(G9) Alternate close working periods with lighter engagement.* The very point of shifting design into use time lies in close contact with the designers and participants at the site of the community. This collaborative design work is, however, resource-intensive, and the necessity of not being there for too long at a time can be turned into a virtue: alternating being there and not can let people find their own ways to engage with the designs and try things out. If effective communication has been established, the community can invite designers to come and spend time, and, just as well, designers may see, for example, from system logs that something novel is happening so it may be a good time to pay a visit.
10. *(G10) Foster ownership.* Aging together seeks to empower the community not just through developing solutions that fit well at some point but also to engage in in-depth competency building work so the community can capably age with the changing media and infrastructure. Designers can advise, help, and elaborate alternatives so that the community can make informed decisions. In terms of collaborative development, this may entail joint discussions and negotiations about which new design directions are pursued and why (cf. Simonsen and Robertson 2013).
11. *(G11) Stay attentive to partial failures, redesigns, and non-use.* As discussed with the taxonomy of active use, people tend to alter design during use and appropriate it differently as well as leave many functionalities unused. These, along with breakdowns and other clear-cut

failures, provide particularly clear cues as to why some solutions do not work and what might be desirable to (re)design (Henderson and Kyng 1991; Hartswood et al. 2002), blending outcome and collaborative design work as happense in (G12) and (G13) as well.

12 *(G12) Embed design across materialities and ways of getting things done.* As we will illustrate through the case of working with Active Seniors below, people can get their jobs done in multiple ways and may prefer social or procedural solutions rather than technical ones. It is thus important to retain an open orientation as to how one might advance the evolution of technology, media, and infrastructure within the community.

13. *(G13) Avoid and anticipate locking in design choices.* Designers almost always hold better command of how different platforms and technologies are developing than the problem owners whose daily life does not revolve around designing. While open and flexible alternatives should typically be favored whenever possible, it is critical to advise the community about the likely pros and cons of different choices over time.

Active Seniors and the Development of Their Information Infrastructure

Let us illustrate the aging together guideposts in practice by recounting the long-term collaboration with the Active Seniors association (Botero 2013; Botero and Hyysalo 2013). The Active Seniors association was set to foster alternative living arrangements for growing old in Finland beyond public and private care homes. Their most throughgoing endeavor was to build the co-housing arrangement Loppukiri, with its related growing-old-together practices (Dahlström & Minkkinen, 2009).[4] Part of the evolution of these practices was an idea of an everyday life information system, which evolved through aging together collaboration premised on long-term active-use and design-in-use strategies.

In terms of social practices involved, the most important ones were those by the Active Seniors association (www.aktiivisetseniorit.fi). Their project was premised on four principles: neighborliness, self-help, community spirit, and open decision-making (Dalström and Minkkinen 2009). The planning and construction of the Loppukiri co-housing building was a six-year process led by the members of the association. To accomplish it, they established working groups for tasks such as fundraising, house planning and interior design, communications, community, and information technology. To prefigure their

[4] Loppukiri translates into English as "last spurt." The plans for the house were being sketched in 2000, and the co-housing collaboration continued through different shapes and funding instruments throughout the construction phase and several years after the seniors moved into their house in 2006.

future practices, rules, and infrastructures, they organized formal events and courses but also parties. Once the Loppukiri was completed, the co-housing community consisted of around 70 people living in 58 apartments and in large common spaces. They organized and cooked a shared regular meal and took care of the upkeep and maintenance of their house through six-week "work shifts" among the residents. Reading circles, yoga sessions, and other social activities are equally important features of the community. They are also actively networked to other co-housing communities and other groups that are building similar projects (Fig 3.10).

Through their activities, Active Seniors have introduced new practices, infrastructures, and service arrangements into Finnish society. In terms of the active-use taxonomy (Fig. 3.4), the project's starting points thus feature locally new designs in ideology and imaginary, community and organization, as well as local settings, meanings, and objects.

In addition to the Active Seniors association, the other key pre-setup for design was the Arabianranta district in which Loppukiri was to be constructed. It was at the time under an intensive regeneration phase and branded as a living lab environment for IT innovation companies with cooperation with local residents and other actors and the University of Art and Design Helsinki (our home base at the time). What this new innovation district might mean in practice was still somewhat elusive, and the Media Lab of the design school wished to develop open-ended collaborative projects to concretize the prospects and scenarios for the district, building on their future housing project and related services (Botero et al. 2013).

Active Seniors contacted Media Lab at the early stage of their project, thus appearing to be a potential match to Media Lab's quest to explore how digital technology could develop in a community that supported design and a capacity to envision its technological needs (Botero and Kommonen 2009). The Active

Fig. 3.10 Members of the Active Seniors community on the Loppukiri terrace balcony (Photo by Sirkka Minkkinen)

Seniors appeared in many respects an ideal partner: they formed an ambitious new community, had their own clear agenda that included design, and were willing to expend long-term effort on future IT infrastructure possibilities even though such interest in information technology was not unilaterally shared across all the members—a key common-sense factor for the Media Lab as well. In turn, the Active Seniors association saw researchers as a potential source for new and useful ideas to further their project and to anchor their project in the district. The parties were also aligned regarding the interest to influence the visions and realization of the innovation district.

This initial part of the collaboration highlights the following aging together guideposts: G1 attending to the practices of existing communities and participants as the point of departure, and G2 clarifying mutual benefits in the collaboration while expecting that they will not be fully shared. This then led to G3, proceeding to verify on this basis whether or not the collaboration could work through a small "access design" intervention: building together a web presence and an intranet for the association, funded through an ongoing research project. In addition to providing added connection between the members who were at the time still living separately, the intranet and internet were thought of as potentially sparking interest in more active computer use among the members by acting as a noticeboard for meetings and events, while the outside facing pages facilitated new member recruitment. The intranet also provided a place to store shared resources such as official documents, albeit it mostly remained in use by the association's board of directors. This access design (G3) helped to understand what the parties could provide in the collaborative endeavor (G2) while anchoring discussions and expectations into something concrete (G4). Already at this point, some partial failures such as selective use of the intranet's possibilities were conductive for the design efforts to follow (G11). In terms of the active-use taxonomy, the access design proceeded mostly "as is" in objects, uses, and settings.

The next phase was to work toward an open agenda for the collaboration (G5). This was pursued through exploratory concept design work, building on participatory design working methods such as facilitated design workshops with the community (Greenbaum and Kyng 1991) paired with codesign techniques like probes and experience prototyping (Gaver et al. 1999; Mattelmäki 2006). The focus points of concept design were decided together and ranged from shared cooking and groceries, care, and security in the co-housing to sharing and coordinating and remembering (Botero et al. 2002; Botero and Kommonen 2009). This supported ideation gave rise to concepts and collaborative design experiments in the shared project but importantly also went beyond outputs that could immediately be turned into design solutions, leaving further "design seeds" for later experiments and shoots for later "indigenous design evolution" among the community members with the resources they had at hand (Björgvinsson and Hillgren 2009).

A good illustration of an indigenous design evolution that emerged was a concept for a video doorbell system. The Active Seniors association's initial concept developed in one of the workshops was that they could have an assigned "on-duty" porter that could view the door video camera and would greet visitors to the house. The "porter" would also receive any calls from residents of the house requiring help. The idea evolved, and by the time the Seniors moved into the co-housing, it was implemented by simply purchasing a mobile phone, which was rotated among the residents of the community. Carrying the phone implied the on-call duty.

The design seeds can be illustrated by some of the knowledge-sharing activities that we experimented with. Some of the senior future residents had very advanced practices for documenting things such as travel stories, gardening tips, recipes, and book recommendations, which served as seeds for some of their collective practices. Based on one of these practices, we prototyped together an audiovisual archive/library for creating and sharing gardening memories and cooking recipes. To act as a design seed, a mock-up was made with existing and new pictures and audio and video footage by some of the members. This prototype itself did not turn into a fixed product but instead seeded other knowledge-sharing practices later on in the house, as we see below. From the point of view of active-use taxonomy, these examples show how existing objects, when paired with active use in meaning, can prompt and also function as new designs. It also conveys the ways in which using can provide handles for further realignment and imagination of new practices.

This early exploration and ideation phase was not just conductive to direct designing but was more broadly important for the members to orient to how life would be once living under the same roof and to hold conversations, map, and envision what it could be comprised of. The concretizations made important aspects of daily life visible to all members along designers, while giving new ideas of how the members could relate to each other and practically do things (G1). This concrete anchoring to future daily practices made it possible to assess technology and infrastructuring possibilities in conjunction with social arrangements and commitments (G2, G12). These scenarios of use and ideas also provided scaffolds for further design and prototyping (G6). The elaboration had instrumental value in providing outputs on which to write funding applications to realize the next steps in the collaboration.

The opportunity to concretize one larger proposal for the Active Seniors association's co-housing arose with the success of a new research project proposal. This effort focused on a shared digital "community calendar" (Lehtimäki and Rajanti 2007). Based on the earlier work and another targeted workshop, the Active Seniors association's working groups elaborated what activities should feature in the calendar and how. The approaching move-in date to Loppukiri led to prioritizing the immediately needed aspects of the envisioned community in the calendar application: communal space usage, sharing, and booking; common meal practices and planning; and ways that neighborly help could be provided. The Active Seniors then wrote the expectations and requirements related to these in the community calendar into a document, and the designers turned them into a paper prototype, which went through several rounds of iterative codesign and, in the process, grew into what the Active Seniors came to call their "Everyday Life Management System." Following this, the design school software development team started to work on a general-purpose framework that could work for the Seniors and be further useful in the future for other communities, with the goal of making it into an open-source software application attractive to a range of communities (Fig. 3.11).

The software came to have six main components, including site and framework for components and use cases; a profile component for information entry; a dining calendar function; a shared resource calendar function for spaces and facilities such as laundry and sauna; a general community calendar for information sharing and events; and personal calendars for all members. A functional prototype of the system was implemented in conjunction to the Active Seniors moving to Loppukiri in 2006. Volunteers tested the system and began to use the calendars for marking events and reservations, as well as for organizing common dinners. The Active Seniors association's IT working group ran "usefulness testing" and ideation to analyze how their practical arrangements met with the system's functions. This resulted in a range of iterations for the rules of the use of the calendar and other components, as well as trials with the alternative interface concepts and terms. Importantly, while the software solution was actively evolving and used by many, others in the community proceeded to coordinate the same issues with a parallel paper-based means they had developed. In turn, the engagement of the designers was intensive during the early implementation and then shifted to monitoring

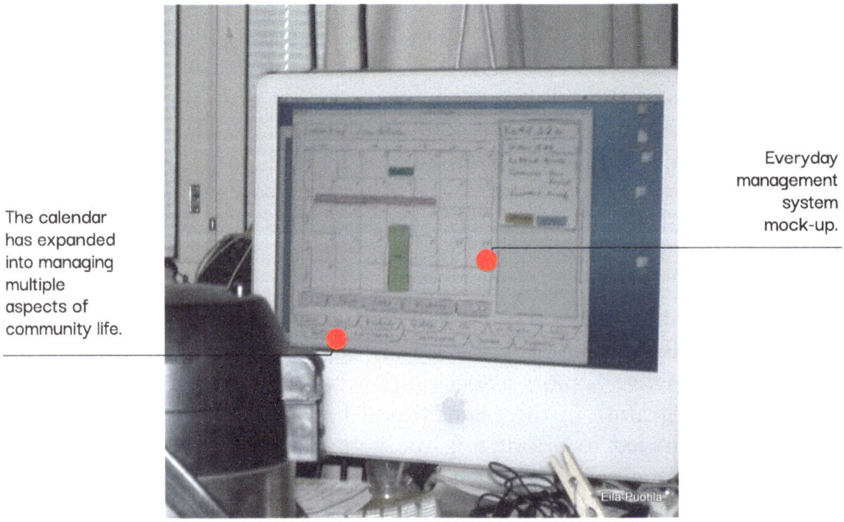

Fig. 3.11 A wireframe prototype for the development of the everyday management system (Photo by Eila Puotila)

the use through the system they had access to and being available upon email requests and meeting invitations.

In terms of aging together guideposts, this period of mixing intense and fluid collaboration could utilize designers first hand and remote presences (G7, G9). In many respects, it was still conducted in a relatively traditional participatory design mode of building upon artifacts and features developed in collaborative workshops in different layers of practice from technology to communal rules to procedures (G8). Crucially, this elaboration was not limited to informing the design and designers to design a truly fitting system but for elaborating the issues for the seniors themselves. The ensuing alteration between close and more sparce design collaboration between the community started to bank on the active use and local designs in objects, uses, and meanings. The community got a chance to appropriate and explore the technology as part of their everyday living practices and then articulate how the various active-use alterations could be further supported. This is important for empowering the community members to take more responsibility for design and technology choices (G9).

The full scope of design-in-use began to be revealed once the community was living together and the everyday practices had stabilized to a point where the potential design space in "everyday life management" could be better assessed and addressed (Botero et al. 2010). This is best exemplified by recounting the evolution of shared meals and the software component related to them. The initial tool dealt with announcing the meals and registering them, but once the community meals were actually held following the move into Loppukiri, an array of further needs emerged. Meal planning and food purchasing for tens of people required information on allergies and other concerns, and this in turn underscored the need for the members to know what would be in a given meal upon registering for it. To this end, the community began to develop its own recipe book based on one they had received as a gift from a Swedish co-housing community. The paper-based recipe book was, however, soon filled with annotations and alterations, making it difficult to follow the different versions, and the paper-based part of the reservation calendar hampered any registering for a meal once outside the building. The seniors consequently asked if it would be possible to implement the recipe book digitally. The design team introduced a fast sketchy version of the recipe book within the everyday life system prototype, reusing the code from another component. There were also changes made to allow easy printing of forms, menus, and calendars. The digital recipe book became a key reason for using the system, even though it continued to hold parallel paper interfaces for meal registering for the members who did not want to use any IT. In terms of active use, this supported local design in objects and uses.

Having the recipe book next spurred on another line of design. At the time the recipe book became available, living together in the house proved to the seniors that they would have many uses for storing and sharing various notes and documents. Thus, some of the active users asked if it was possible to make a "copy" of the recipe book, which the designers provided. With a copy of the recipe book, the seniors could devise workarounds and make the copy behave like a community "notice board." They altered the titles to get them to appear in a preferred order (e.g., by adding numbers or other symbols before titles to manipulate their sorting) and "misused" other recipe fields to create news-like content. The recipe book thus seeded indigenous design evolution within the community (Fischer and Ostwald 2002). Further on, once the Seniors contacted the designers with their hacked local design (active use in object) and how it was to be used (local design-in-uses), a joint workshop was held to explore how the changes could be incorporated on a more robust basis. The software developers further generified the solution into "dynamic information containers" and exemplified this with a new recipe book, new notice board, and document storage spaces. Again, the solution was built purposefully as an inviting design seed to foster further uptake, and indeed, the Active Seniors responded by creating several new containers, for instance, for announcements, documenting events and activities, and swapping and loaning things. The generic design further made it possible to link the containers in different ways to build more advanced structures, but the software developers did not have the resources to make an adequate user interface for this, and these possibilities thus remained underused. In terms of aging together guideposts, we thus see the interplay between being in the community (G7), empowering them to engage in active use (G10), and alternating intensive and hands-off engagements between designers and users (G9).

A final aspect we wish to highlight from the Active Seniors association's engagement concerns the interplay between technology design, daily practices, and infrastructuring, which by no means proceeds through technological solutions being the ones preferred. Despite the careful collaborative design of calendars for sharing spaces and facilities, their usage also featured challenges. At the time in the mid-2000s, it was unfeasible to install interactive IT interfaces next to all the spaces and facilities, and the booking and information about the bookings was not available in real time in the actual space, and there were still community members who did not use IT at all. It proved tricky to keep both paper and online calendars up to date, and, at some point, paper became preferred over the digital version.

Just as well, some of the common daily operations in using the system turned out to work better through social agreements rather than by changing the code. It was common that neighbors would help each other and take actions on others' behalf, yet the software architecture did not support such collective usage, as it demanded individual logins and passwords. This could be worked around to some extent by active use in, for instance, agreeing to set passwords so that the people who would commonly use the system on behalf of another would easily guess the password.

These examples underscore the need for versatility not only in alternating the intensity of design collaboration (G9) and "who does the design" (G11) but also the mode by which design solutions are pursued from technical to procedural to social agreements and rules (G12). This further called for efforts to foster preferred co-evolution of daily practices and technology across many layers including infrastructural long-term strategies (G13) (Hillgren et. al. 2011; Karasti et al. 2019). In terms of active-use taxonomy, the aging together case underscores how different active-use engagements can intertwine with professional design efforts once the design departs from the practices of the participants involved and continues into the design-in-use time. Moreover, the case shows how different active-use engagements tend to feed into each other, changing the focus of design engagement from objects to uses and from new local meanings to altering the local settings (for visually making such mappings, see Botero and Saad-Sulonen 2022.) To visually clarify how this intertwinement can look like in terms of the active-use taxonomy, we add a gestural movement layer (Agid and Akama 2018) to our static figure (Fig. 3.4) to trace the trajectories (Hyysalo 2010; Botero and Saad-Sulonen 2022) of these engagements. In Fig. 3.12, the Active Seniors project's active use intertwines with the design-in-use example of the video porter experiment. Moreover, to illustrate how different active-use engagements tend to feed into each other, changing the focus of design engagement from objects to uses and from new local meanings to altering the local settings and sometimes imaginaries and arenas, Fig. 3.13 maps a couple of points in the evolutionary trajectory of the Active Seniors association's everyday life management system. These should underline the dynamism at play in making design and active-use movements and connections across the whole range.

Chapter Conclusions

Active use and design-in-use strategies underscore that the extent of "play time" of design participation affects attainable outcomes in important ways. Keeping codesign "on" throughout concept design, launch, redesigns, and further releases can benefit from means such as setting up codesign workshops, creating and making visible design seeds, employing observational techniques (and so on). A more throughgoing orientational shift toward design participation is, however, also required—taking these commonly used ways of working not as vehicles for arriving at well-fitting one-time design but as means for fostering the evolution of practices and their technological mediation during use time. Such an approach recognizes that in this co-evolution, it is the community of practice and its daily practicing that is the primary yardstick for success, not the technical prowess or design elegance of the solutions. The focus on co-evolution means that the social practices, design solutions, and

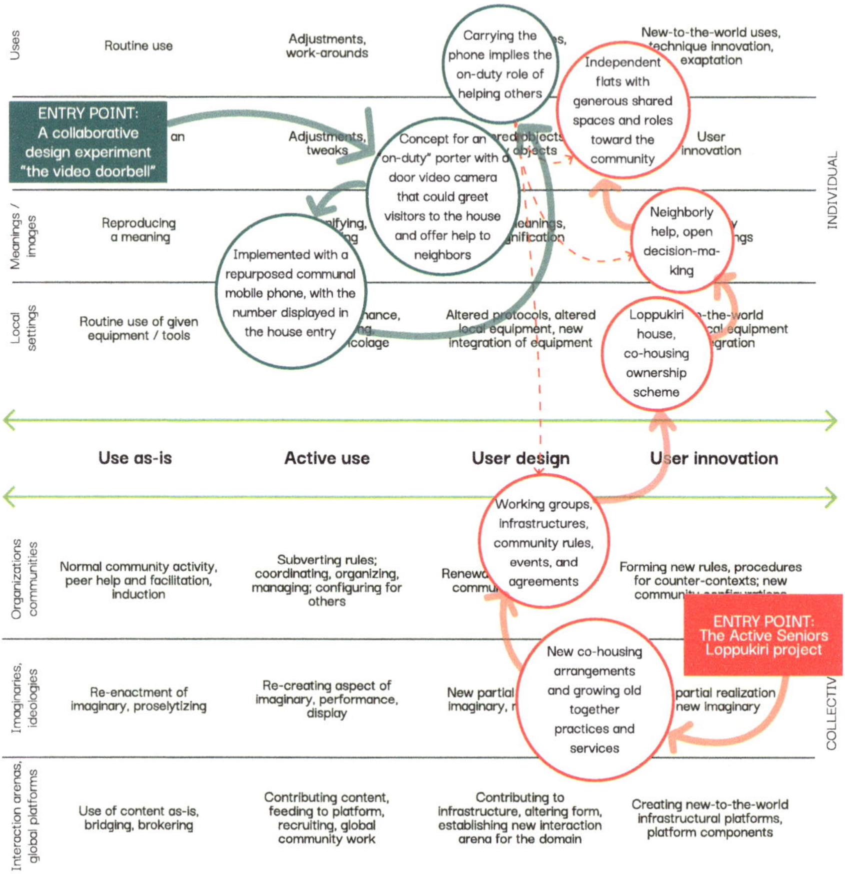

Fig. 3.12 Aging together trajectory example 1. The Active Seniors project's active use (in green) intertwines with design-in-use around a small collaborative design experiment (in red). The dotted lines are centering points (in time)

infrastructuring need to "age together"—that is, evolve into practice-affording arrangements over time and be "corealized," as in reasoned and worked on together (cf. Hyysalo and Botero 2013; Hartswood et al. 2002).

While in users' design collectives this at best happens naturally, in settings where professional designers engage, seek to build, or foster design-in-use strategies, they require new sensitivities and ways of proceeding. This concerns both exploratory and academic projects as well as the ways of pursuing human-centered design in contexts where, for example, digital services are launched and developed with minimal viable product strategies (Johnson 2013).

Such an extended timeframe evades attempts to think of design participation as a matter of following a single particular design method, which is

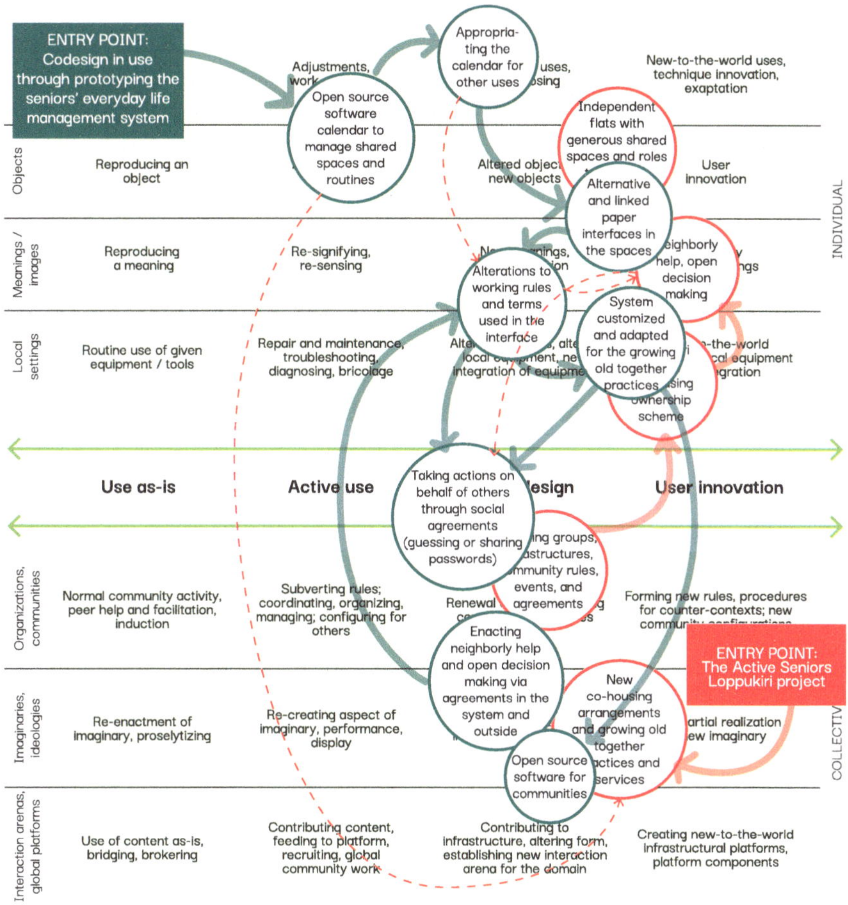

Fig. 3.13 Aging together trajectory example 2. The Active Seniors project's active use (in green) intertwines with design-in-use intervention and infrastructuring around an "everyday life management system" (in red). The dotted lines are centering points (in time)

common in many UX and codesign approaches. The aging together orientation challenges design participation to attend to a repertoire of long-term design engagement that should be more aware of the converging, parallel, and complementary aspects of design-in-use engagements. We believe the strategies introduced here are a resource from where design practitioners (and design collectives) can draw on when they reflexively attend the particularities of the given design engagement and try out what may and may not work there (Schön 1983).

As the taxonomy of active use indicates, many people are quite adept at engaging with different aspects of design in their own domains of practice. It also means that, at times, there could be demands for finding resources and tools, mobilizing them, and creating conditions to support the collective unfolding of a truly collaborative design space (Botero 2013). Aging together shows how design-in-use thrives through moving between different active-use engagements and how those often set preconditions and open possibilities to further active engagement (Botero and Saad-Sulonen 2022). Equally, our example points to bottlenecks that professional designers can address to augment users' design engagements and the expansion of the design space for the members of a collective. Hence, what is at stake with design-in-use strategies is not only a thorough engagement with the designed products/services/media, it is also mindful reconfiguration of social practices and infrastructuring the settings and platforms involved, while mindfully—but opportunistically—exploring the full scope of the active-use taxonomy (Karasti et al. 2020; Bücher et al. 2009).

Shifting design into use time and taking use of active-use phenomena are arguably at its strongest whenever technology architecture is modular enough to allow quick alterations and users practices are rapidly evolving, such as digital services launched as "perpetual beta" (see Chapter 6). As other design-in-use methodologies have underscored, it also makes sense in high dependability and high value settings where work-affording technology is of paramount importance (Voss et al. 2009). Yet the aging together approach and the Active Seniors example shows that with designers' involvement, such evolutionary design-in-strategies can also well work with everyday people who are not deeply knowledgeable of the particular technology or media that is being designed. While the aging together engagement with Active Seniors was, at the time, made possible through research grants, similar engagement with communities of practice can be realized with fewer resources, banking on the long-term evolution and alteration between intensive and hands-off relations between designers and communities.

CHAPTER 4

Design Participation in Steering Wide Sectoral Change

Introduction: Expanding Design Participation to Steering Sector-Wide Systemic Changes

This chapter outlines what design participation can offer for steering sector-wide systemic changes.[1] We begin by outlining the need for such transformations and the dynamics of sociotechnical change that render such changes difficult and arduous to achieve. We then introduce a mid-range transition pathway toolkit (MTPT) and arenas as one avenue of how design participation can help foster sector-wide changes through helping diverse actors in envisioning and building pathways toward a more sustainable future. After this, we dive in to consider the reorientations that may be required from designers and other actors who seek to foster such transformative changes.

The design profession has expanded the scope of design objects throughout its history. Moves from industrial design to interaction design, from information design to service design, from ergonomics to user-centered design, and from codesign to open design have all been associated with new types of objects, competences, processes, and relationships with the design clients and users (Valtonen 2007; Abel et al. 2011).

In addition to extending design activities into use time (see Chapter 3), a recent key development in the field of design has been the expansion of design

[1] This chapter builds on and extends two articles that we published under CC-BY–NC–ND 4.0 licence. Hyysalo, Sampsa, Marttila, Tatu, Perikangas, Sofi & Auvinen, Karoliina. (2019) Codesign for transitions governance: A mid-range pathway creation toolset for accelerating sociotechnical change. *Design Studies* 63, 181–203; Hyysalo, Sampsa, Marttila, Tatu, Perikangas, Sofi & Auvinen, Karoliina. (2019) Intermediate codesigning in transitions governance: Catalysing and channeling participant action. *Design Journal*. All the coauthors have granted Palgrave (Springer) the right to use the text in this commercial open access book.

S. Hyysalo, *Design Participation*,
https://doi.org/10.1007/978-3-031-95437-5_4

activities into long-term transition processes toward, for instance, low-carbon energy systems and circular economies (see, e.g., Ceschin and Gaziulusoy 2016; Irwin et al. 2015; Hyysalo, Marttila et al. 2019; Hyysalo, Perikangas et al. 2019; Marttila et al. 2023; Lähteenoja 2024; Auvinen 2024).

Transition governance is an increasingly central area of concern among policy, business, and design practitioners, as well as among academic researchers (Köhler et al. 2019). Such society-wide long-term changes are not "designable" per se, as they result from hundreds of intertwined actions that span regulation, technology development, altered consumer practices, taxation, and new business creation (and so on) (Geels 2004; Köhler et al. 2019). At the same time, there is plenty that can be designed for furthering sociotechnical transitions (see, e.g., Ceschin and Gaziulusoy 2016; Jalas et al. 2017).

An important strand of this work concerns the use of design participation to better connect the relevant actors needed for strategic action in bringing about societal change in liberal democracies: decision-makers, experts, civil servants, citizens, non-governmental organizations (NGOs), and business leaders (to name a few). Codesign and participatory design have a long tradition of creating contexts, tools, and procedures through which such diverse stakeholders can better deliberate and learn from each other (see, e.g., Schuler and Namioka 1993; Bødker et al. 2004; Botero 2013). However, when used for catalyzing social change and sustainable alternatives, the outcomes of design participation may not result in further designs but in more open-ended, wider, and loosely defined outcomes—such as envisioning new organizations, regulations, and altered everyday practices—and the schematics of how these are tied together in fostering change (see, e.g., Jalas et al. 2018; Gaziulusoy and Ryan 2017a, b; Hyysalo, Marttila et al. 2019). Such pursuit is well warranted, as achieving thoroughgoing transitional shifts requires anticipatory action, societal experimentation, and policy changes (Geels and Schot 2007; Köhler et al. 2019). To understand why, we need to make a quick dive into what is involved in sociotechnical transitions toward sustainability.

Why Is System-Wide Change Difficult?

The need for thoroughgoing system transitions has become urgent in several areas, such as energy, housing, and transport, in which climate change exerts pressure to replace the fossil-fuel-based systems (IEA 2023; Köhler et al. 2019). Such transitions affect society widely, going beyond energy production and consumption to include zoning, land use, taxation, mobility solutions, consumer practices, and so on.

The steering and governance of such long-term systemic transitions has been investigated since the late 1990s in several multidisciplinary lines of research. It has become evident that single alternative design solutions cannot compete against the inertia created by "sociotechnical regimes" built over decades by interlinkages in production technologies, investment patterns, scientific bases, institutions and policies, established market mechanisms, and the culture of consumption (Geels 2004; Geels and Schot 2007).

At the heart of difficulties to achieve sociotechnical change lies a phenomenon called *path dependency*. Path dependency results from historically cumulated *learning effects*, *network effects*, and *sunk investments* that have shaped the socio-techno-economic environment in the course of history to the effect that technically or societally inferior solutions continue to be favored against the alternatives (David 1985, 2007). A classic example is the QWERTY keyboard, which a gross majority of people use in their typing despite it being slower to learn than alternatives that are just as fast for typing (e.g., alphabetically organized keyboard) and slower to type than keyboards better optimized for most common letters (e.g., Dvorsak keyboard) and syllabi (dicta machines). QWERTY was originally introduced by the Remington company in the nineteenth century as a solution for their mechanical typewriter to avoid the punching arms becoming tangled when their sales reps wrote "typewriter" by only using the top row of letters. While only a sales gimmick, Remington's design made the many who adopted their invention use this keyboard layout, making the switch and relearning costly for the touch typists, thus fortifying the *learning effects*. As the large administrative organizations of the time required large cohorts of touch typists, and typists themselves preferred a transferable skill, the QWERTY layout became the preferred de facto standard to which competing manufacturers also shifted, inducing ever-widening *network effects*. Learning to type with alternative or multiple layouts became increasingly disfavored, and for the organizations, the *sunk investments* made into stocks of QWERTY machines and QWERTY skilled typists acted as a further barrier to switching to the alternatives. Once Olivetti introduced the first electric typewriter in the mid-twentieth century, the original typewriter arm rationale disappeared, yet QWERTY has continued to reign as a layout even as personal computers have long allowed changing the layout in a manner of minutes (David 1985).

QWERTY's path dependencies are a good entry point to transitions, as it has gained and lost many important *connections* that have contributed to its continued dominance. In other path-dependent processes, these connections have become importantly further fortified. A private petroleum car exemplifies some of the important ones. Soon after its introduction into the horse-carriage-dominated world of the nineteenth century, some cities required a red-flag-carrying runner to lead these noisy dangerous machines (Kivimaa et al. 2019b). This infrastructural dwarfing of their potential gradually reversed as the popularity of the car grew and the learning effects cumulated in the associated technology, science, and operation. By the 1930s, the first motorways banned other transit, and new peri-urban and suburban forms of cityscapes emerged, reliant on the car's ability to connect the areas to cities and towns. These *infrastructural interdependencies* became gradually fortified through transit planning and regulation with massive investments into paved roads, parking lots, conveniently placed gas stations, gasoline refining and delivery (and so on) and, just as importantly, through urban planning principles and regulations that dictated the urban form amenable for the private car (Hoogma et al. 2002). Add to this consumption patterns and cultural norms on how to access workplaces, recreational activities, summer houses, and the like, and the functioning car transit *sociotechnical regime* has come to feature high *society-wide criticality of operation*—disturb the car driving and send wide repercussions across society. Just as an example, activists planning to block a major highway in England were recently sued for terrorism. Given the predictability of such an environment and continued customer demand for cars and all the related goods, producers have been able to benefit from economies of scale and invested in extremely efficient production lines and long production runs that have pushed the unit costs down to a fraction. Such a state of affairs has been described as a *sociotechnical lock-in,* in which it has been exceedingly difficult and slow to introduce any alternatives, as all alternatives need to play by the system-wide alignment and rule sets cast for the petroleum private car (Hoogma et al. 2002; Kanger et al. 2018; Hesselgren et al. 2016).

Against this backdrop, the strategic niche management research shows that, to have a chance of substituting or reconfiguring sociotechnical regimes, the alternatives need to first mature in "niches" where innovations can develop without facing the full impact of existing market conditions. The support measures that nurture, empower, and shield niche innovations, such as feed-in tariffs and funding for experiments and pilots, can be gradually removed once niche innovations mature and their production processes become cost-competitive (Hoogma et al. 2002; Kivisaari et al. 2004; Smith and Raven 2012; Köhler et al. 2019). At the same time, destabilizing measures such as new policies and policy mixes can be set in place to destabilize the incumbent regimes and make room for change. Typical examples are carbon pollution pricing and legislative carbon-level bans (Kivimaa and Kern 2016; Weber and Rohracher 2012).

The long-term transitions have also been found to require direction and guidance beyond the political cycle of elections. To this end, *transition management* (TM) originated in the Netherlands in the 2000s, and since then, a methodology for setting long-term visions, creating change pathways, and identifying experiments to start such pathways of change has been developed (Kemp et al. 2007; Loorbach and Rotmans 2010; Frantzeskaki et al. 2014). The aim of TM is to create spaces for searching, learning, and experimenting on the transformation of the current system with "frontrunner" stakeholders (Kemp et al. 2007; Loorbach and Rotmans 2010). Within TM, the construction of pathways of change to meet a long-term vision and specific transition goals as well as corresponding experiments that can lead toward change are further emphasized (Loorbach and Rotmans 2010). The identification of challenges, vision building, and the construction of pathways of change take place in transition arenas (TAs)—series of workshops conducted with diverse groups of pioneers. TM and TA processes have been geared toward a long-term transition focus of 40–80 years, and the means of creating scenarios and pathways have reflected this, remaining relatively broadscale, connected to present concerns with the identification of "immediate actions" (Frantzeskaki et al. 2017; Roorda et al. 2012). The schematic overview of TM is as follows (Loorbach and Rotmans 2010):

1. Establishing a transition arena (or arenas)
2. Developing a common vision
3. Developing pathways through "backcasting" techniques
4. Experimenting with pathway options
5. Monitoring, evaluating, and revising pathways and experiments.

In recent years, design for sustainability transitions has entered onto the transition research and governance scene, particularly regarding experimentation with new solutions and in improving the means for future envisioning. Design research has generated experiential future scenarios and change pathways (Gaziulusoy and Ryan 2017a, b) and has pursued sustained local experimentation engagements aimed toward low-carbon transitions, drawing from community design and practice theory (Jalas et al. 2017; Manzini and Rizzo 2011; Mok and Hyysalo 2018). Design agendas have also been proposed that resonate with designing for sustainability transitions, such as transition design (Irwin 2015; Irwin et al. 2015), designing for environmentally sustainable social innovation (Jégou and Manzini 2008; Manzini 2014), designing with intent for sustainable behavioral change (Lockton 2017), and designing for one-planet lifestyles (Lettenmeier 2018). The potential of design has begun to be noted by other disciplines, such as codesign being seen as a contributing field to TM (Ferguson et al. 2013). But to deliver on the promise, design research needs to build more convincing real-life projects and show the value that different types of design research can provide in designing for long-term sociotechnical transitions. The design participation efforts described in this chapter are part of this effort, exemplifying what codesign for sustainable transition can offer.

Mid-Range Transition Arenas and the Pathway Toolset

The increasing urgency to accelerate sustainability transitions calls for better means to address the mid-range dynamics taking place in the 5–15 years ahead, as it forms the most relevant policy timeframe considering the large gap to e.g., carbon neutrality and the mitigation of biodiversity loss around 2040–2050 and the current situation (IPCC 2018; IEA 2023; IPBES, 2019). Shifting from long-range to mid-range visions and pathways, however, makes goals and pathway steps more concrete and potentially more difficult to reach an agreement amongst the stakeholders. Furthermore, in the TM manuals, the pathway construction process in TAs builds on deliberation among participants, which the analysts then turn into coherent pathway depictions (Ferguson et al. 2013; Roorda et al. 2012). In our view, this potentially underplays the potential that the participants have as *the* domain experts capable of directly constructing the pathways. In the terminology of *International Association of Participation*, there appears to be room to move from informing and consulting modes of participation (see Chapter 5) to participant involvement and collaboration (IAP, 2023).

These participation opportunities in mind, the means by which the participants work in the TA processes called for improved intermediate designs (Eriksen, 2012; Eriksen et al. 2014; Vaajakallio 2012) that could aid more hands-on multi-actor engagement in path formation in fast-paced workshops (Hyysalo, Marttila et al. 2019; Hyysalo, Perikangas et al. 2019). Our response has been the design, deployment, and make of several iterations of the Mid-Range Transition Pathway Creation Toolset (MTPT) and the associated mid-range TA process.

The MTPT, at the time of writing, comes in two basic versions and several variations. The offline version is comprised of a large metallic board on which workshop participants can move pre-formed magnetic elements; procedures for progressing in the work; facilitator and participant guides; and a digital environment that allows the distributed refinement of pathways between collocated sessions. The online version uses a slightly simplified notation system ran on top of the Miro™ platform and Zoom conference to allow for a spatially distributed process. The toolset allows participants to directly construct pathways of change; proceduralizes the vision and goal-setting processes to reach an agreement over potentially conflicting targets; proceeds through anchoring the vision and goal-setting phases in the typically existing public visions in a domain area and country context; provides background memos and elaboration of already running experiments in the change domain as starting points for the arena process; and features additional commentary options for wider civic society involvement (Hyysalo, Marttila et al. 2019; Hyysalo, Perikangas et al. 2019; Hyysalo, Lukkarinen et al. 2019; Marttila et al. 2023; Lähteenoja et al. 2023, 2024). To be able to enroll busy activists and industry change makers and equally busy public-sector actors, the TA process has been compacted from the 10–14 full days suggested by some TM manuals (e.g., Ferguson et al. 2012; Roorda 2013) to a series of 3–6 workshops that last 3–6 hours. In between the workshops, participants comment on refined results from the previous workshop in the closed website of the arena. A typical six-workshop series would be comprised of the following, taking place in 2–4-week intervals:

Workshop 1. Identifying the drivers, challenges, and contingencies for transition

Workshop 2. Elaboration of the vision and mid-range interim transition goals.

Workshop 3. Formation of pathways, part 1.

Workshop 4. Formation of pathways, part 2.

Workshop 5. Immediate actions for launching the pathways.

Workshop 6. Completing the results and iterating the final report.

At the heart of MTPT is a set of predefined forms and categories, which are used in constructing change pathways and were designed to give sufficient visibility to both content and form for all the participants during the process and to both ease the movement of the elements and transform the pathway during the pathway construction.

The primary elements of the pathway creation system are the "pathway step" and "pathway-step action" elements. Both have the same structure: upmost, the designator of the form (such as *pathway step* or *investment*); then four rows for describing the step content; followed by timing (in years), the key actor(s), and the scale(s) that this element concerns—a national issue, a regional issue, on the suburb/village scale, or concerning individual buildings and consumers (Fig. 4.1). To differentiate the elements, a combination of distinctive symbol, text, and color is used for each.

The pathway-step action elements concretize how each pathway step can be realized or facilitated. These are specific to the domain area in question and thus feature some variation. When working with energy transition, the pathway-step action elements are energy production, business, end consumption, regulation, investment, technology, pilot, and other (Fig. 4.2, left-hand side). We also designed a set of organizer elements to guide the work. "Fact elements" are used to render visible key milestones and facts about the pathway (see more below), and the question mark, exclamation point, and quotation marks are used to point out missing or insufficient pathway steps regarding change targets, critically important areas, and needs for new research, respectively, with the aim of focusing participant attention on these areas (Fig. 4.2, right-hand side). The choice of hexagon-shaped elements, descriptive labels, and color coding was based on their common use in countless board games and ideation systems (Hodgson 1992).

The interrelations between elements can be clarified with magnetic arrows to show how one pathway step leads to another (arrows allow writing on

Fig. 4.1 A pathway-step element and an example of a filled-in pathway step. (Hyysalo, Marttila et al. 2019)

Fig. 4.2 On the left are pathway-step action elements: energy production, business, end consumption, regulation, investment, other, technology, and pilot. On the right are organizer elements: a fact, an attention marker, a missing action marker, and a research marker. (Hyysalo, Marttila et al. 2019)

them, as many interrelations are not strictly causal). Once the pathway is completed on the board, it is rendered digitally, which allows further commentary, cleaning, and the conversion of all content into full sentences that are understandable to those beyond the path creation participants (see Fig. 4.3 for a completed pathway).

Prior to the pathway construction, the organizers produce a 4–6-page information package for the participants, which is related to the pathway and domain that condenses the current state, the envisioned pathway goal the participants have elaborated on in the preceding workshop, basic facts in the change domain, known challenges, and some of the evident basic calculations such as what would be the carbon reduction from different likely measures to be considered. The info package cannot level participant domain expertise differences (which can be huge!), but our experience is that it can importantly spread the knowledge that the diverse participants have regarding different aspects of the change domain and, just as importantly, take the pain out of remembering quantities such as current and prospected capacities of, for example, energy generation and consumption during the pathway sessions.

The information in the package is also partially rendered visible on the board by placing key facts and pilots tentatively on the board as prefilled fact and pilot elements (see the green and blue elements in Fig. 4.3). Initially, a

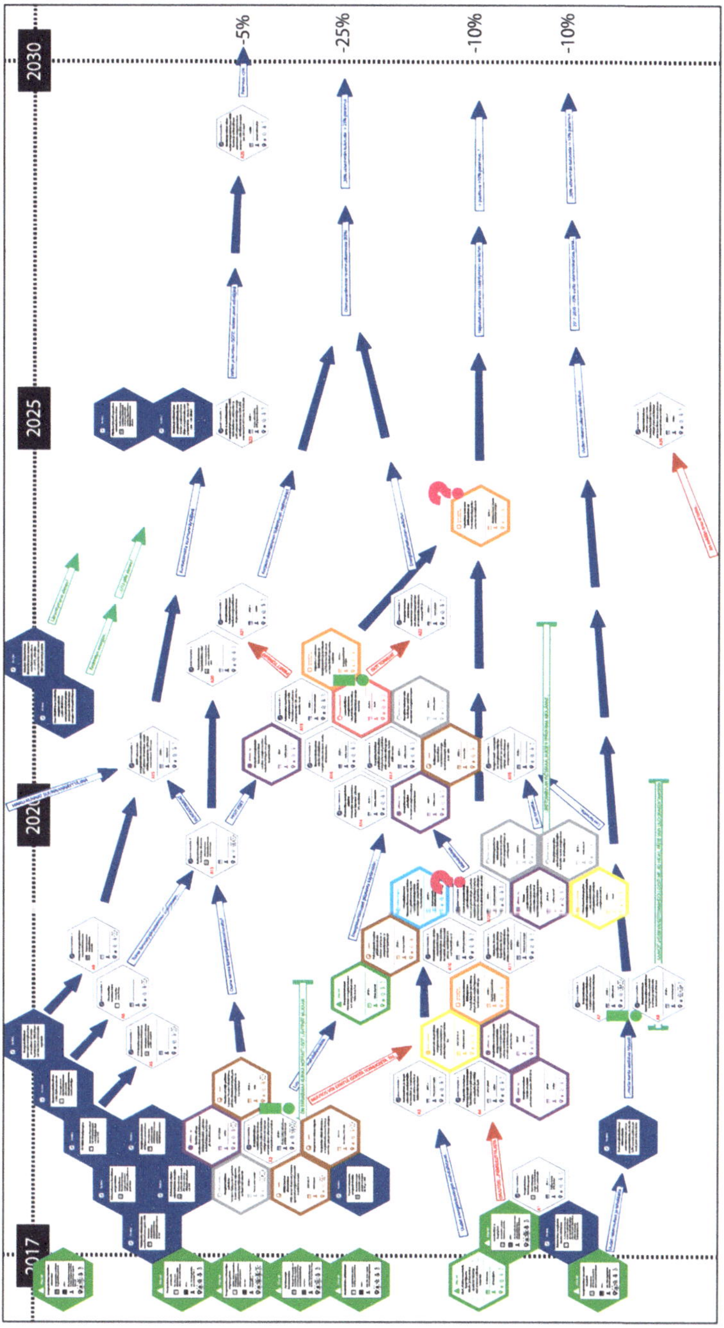

Fig. 4.3 A digitized path from 2017 TA process on halving Finnish building stocks net-energy use by 2030

data-derived “persona” (Cooper 2004) was attached to focus attention on the concrete people who need to take the change actions—such as a Finnish family living in 2030, implicated by the pathway—but this was found to narrow the attention too much because the pathways invariably implicated dozens of different key actor groups.

From the starting position, the participants begin by discussing the target and pathway on a general level. The facilitator urges them to write down their thoughts about pathway step elements whenever an obvious step is identified. As steps accumulate, discussions begin to include their interrelations and potential missing steps. In all the paths created thus far, the elements were rearranged several times and sub-pathways emerged, either from the onset or through branching. At some point, the deliberation tends to veer toward considerations of whether each step is needed, whether some steps are realistically attainable, and whether all the steps in all the sub-pathways together amount to sufficient change regarding the transition goal (Fig. 4.4).

Once the main pathway steps have found a steady and agreed-on form, the participants move on to identifying the most important and most crucial steps as well as what blocking points may occur in the pathway. This constitutes the first phase in the pathway construction. At this point, the first documentation round happens through asking participants to explain to the video camera the key features of the pathway and new insights they gained during the path construction.

The second phase of the pathway creation process is a more detailed examination of the most important steps. The actions needed to realize each pathway step (e.g., technology development, regulation, changes in consumer behavior, pilots, investments; see Figs. 4.2 and 4.5) are discussed and marked down. At this point, it is common that some pathway steps merge, and some

Fig. 4.4 Pathway creation in its early stages

Fig. 4.5 An example of a pathway step for which the facilitating actions have been explored in detail. (Hyysalo, Marttila et al. 2019)

new steps are added, or the status of a pathway step is changed to be one of action needed for realizing another pathway step. At the end of the second phase, the participants video the concretizations detailed to ensure that the ideas written down on the cards are sufficiently explained.

The third phase of the process progresses into uncertainties and contingencies. At this point, the facilitator changes from blue markers and blue arrows to green ones and adds in probability markers of varying lengths. The participants then go through each step and examine the likelihood of the steps; can they occur sooner or later and how uncertain are they? The overall uncertainty factors are already identified in the second workshop of the series and can now be used to gage the uncertainties related to specific pathways. The participants then add potential contingency responses, which are marked with green arrows and green-stickered and green-text pathway steps. The final phase of the pathway creation is considering the alternative, mutually exclusive change pathways to the original pathway. These are identified with red-stickered steps, red texts, and red arrows. This step is done last because alternative paths typically require rearranging the original paths; thus, the originals must have been first documented without interference from mutually exclusive paths or steps.

Once the entire pathway is complete, it is digitized and uploaded to the password-protected support website of the transition implementation arena, which gives the participants the opportunity to further comment and refine the pathways after the workshop. If pathway construction is spread to multiple workshops, incomplete pathways can be digitized and shared in the platform to allow commentary between sessions.

Pathway creation relies on following the procedures, facilitators' assistance, and their actions to keep both the participant discussion and path construction actions on track. To aid this, both detailed participant instructions and facilitator instructions exist, along with a guide for how to transfer the physical board's state into a digitized environment in a unified way. It has been found useful to use a facilitator and a notetaker who both participate in digitizing the contents, but we have also run the workshops with a single facilitator.

The online-only version of MTPT has been developed on top of Miro™ group ideation software by creating the same hexagon grid and somewhat simplified pathway elements by removing the information fields on geographic scope and time (which is implicated by the timeline as well) and shortening the information on actors' implications into abbreviations written at the end of the substance text. The online interaction tends to be less dynamic and slower—participants typically take their turns to speak and elaborate on pathway elements instead of complementing each other's entries—but consistently nonetheless deemed sufficient by participants and organizers. In addition to easier logistics, the online-only version has the advantage of resulting in readily typed text, which reduces post-workshop processing time to about half of the physical digital format (Fig. 4.6).

The uptake of the mid-range arenas and MTPT tools have been promising since their initial launch in 2017 to the time of writing in 2025. Altogether,

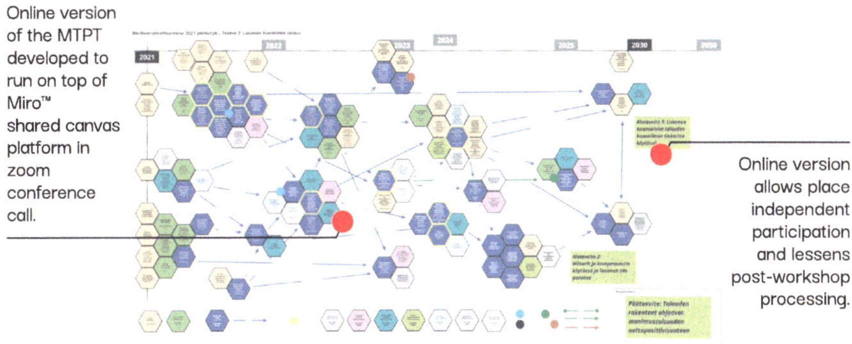

Fig. 4.6 Online version of the MTPT pathway built on top of the Miro™ collaboration platform

14 mid-range arenas have been organized, and several more are in preparation. This is a significant amount considering that most arenas last for several months. The arenas have been targeted at different scopes of change and societal sectors and ran with some variation in orientation and the way they have been organized.

Some arenas have had a regional scope. For example, two arenas have targeted the blue bioeconomy in different regions (Valve et al. 2019a, b; 2023). And some arenas have focused on a particular city, such as one arena's focus on carbon reductions in mobility in Vaasa (Enell-Nilsson et al. 2019) and another's on a low-carbon district in Tampere.

Many arenas have been conducted at a national scope. The first arena was an overall energy transition arena (Hyysalo et al. 2018, 2019a, b, c), followed by a citizen energy arena (Lukkarinen et al. 2020, 2023; Lähteenoja et al. 2022), and further followed by a suburban low-carbon housing arena (Lukkarinen et al. 2023a) and peri-urban low-carbon housing arena (Lukkarinen et al. 2023b). Similarly, arenas in sustainable packaging (Varho et al. 2023) and plastic recycling proceeded with a national focus.

Another aspect of the development of mid-range arenas has been the direct commissions from the Finnish government to aid policy processes. The first of these focused on the new Finnish biodiversity strategy in 2020, commissioned by the Ministry of Environment (Hyysalo, Lähteenoja et al. 2022), and the second focused on the national implementation of the United Nations Agenda 2030, commissioned by the Prime Minister's office (Lähteenoja et al. 2022; Lähteenoja 2024). These two commissioned arenas differ considerably from others and form a "policy edition" of mid-range arenas (Lähteenoja 2024). The arenas were part of formal policy preparation processes; the participant groups were large, between 50–80 people; and the participants were no longer volunteer front-runners as in other TMs but came by the request of the Ministries. And, albeit the participants were still diverse, administrative and lobby organizations were proportionally more represented than industry and civil society actors. The tie-in to formal policy processes led to more formal positioning by some of the actors and actors ensuring that their core interests were represented and not jeopardized in the final reporting. Still, many participants reported that the arena processes were able to push change forward and find common ground on some of the (quite entrenched) issues between interest groups (Lähteenoja 2024).

In terms of impacts, the arenas have varied, with the most notable impacts being the direct uptake of some of the key proposals from the first energy transition arena to the Finnish government program, and many envisioned actions later having become implemented. The commissioned arenas have shaped the policy preparation processes, but it is still too early to assess their overall role, as these policy processes have had very long time spans. Other arenas, such as those targeted at citizen energy, have been conducive to get new actions started and for the actors to begin to form coalitions in fostering the changes in this sector (Lähteenoja et al. 2022; Lukkarinen et al. 2022). The "impacts" are never this unidirectional and merit further concentrated research and conçeptualization, but the above should give some indication about the substantive effects in steering transformational sociotechnical change that design participation can have.

The Intermediate Codesign of a Mid-Range Pathway Formation System

The strategic-level codesign for transitions introduces some specifics for design. These are worth recounting and likely prove helpful for others seeking to foster future-oriented design participation and coproduction of knowledge among diverse stakeholders. The following are baseline observations of what is involved and what is emphasized in comparison to "conventional" design participation in terms of its intermediate design work (we leave aside other design participation work types here):

- *Entangled design objects*: Codesigning for transitions blends the design of physical elements, the designing of and for interactions among participants, procedures, principles, facilitation, and documentation. None of these can really be designed in isolation and without testing how their interactions play out.
- *Multiple and potentially conflicting "goods" and design goals* are common in most design, but the goal conflicts are particularly salient in intermediate designing for transitions, which is to feed into outcomes and engage the participants in the process, seeking to foster (sometimes multiple) preferred modes of interactions (such as high-quality deliberation, complementary knowledge coproduction, and effective pathway construction).

- *Distribution of agency between designers and participants* is a continuous and core consideration. In deliberative settings where the prime people to act on the results are the participants, designers need to extensively invest in thinking about what results would be actionable for the participants and in which form they would then need to be. At the same time, the designers need to consider how the ownership of the process can become adequately distributed: which aspects of the deliberation and outcome production can be preset prior to the process, and which are best left to participants during the process?
- *Sufficient knowledge of the target domain and participating actors* is key to achieving intermediate design that works for the task and for the participants. This may require multidisciplinary collaboration, such as in the initial creation of MTPT, where the specific area of the energy system warranted involving a specialist as part of the design team and a whole array of trusted energy system experts as testers prior to the first launch.
- *Reducing the complexity to minimal requisites in representing the target system*: High simplification would allow easy learnability, easier workshop design, etc. but would also run the risk of producing trivial or shallow results. For instance, as we learn later in this chapter about the process of developing the MTPT, the mid-range energy transition pathways featured extra steps that were more complex, more numerous, and more difficult to construct (even for the experts) than had been anticipated. This resulted in continued efforts to cater to this complexity in the intermediate design, while keeping it from overwhelming the process: first moving into depictions of individual pathways instead clusters of pathways, then reducing the systematicity in resilience analysis, sequencing the pathway construction process into four distinct phases, and finally moving to pathway interrelation analyses only after the workshops.
- *Tradeoffs between principles and mundane considerations* are also part of intermediate designing in codesign for changing sociotechnical systems. For instance, in the MTPT system, ideals of high-quality deliberation and rigor in resilience analysis not only competed for time with each other but also with the attainable time from participants, as well as the manageable number of people per pathway board, facilitator effort, and resources regarding how many boards can be worked on simultaneously and the effort and time participants needed to expend on learning alternative notations and procedures.

Overall, the nature of design under these conditions can be best characterized as *designing to channel participant action*. The intermediate design effort kept on balancing between design choices that would give decision-making power and ownership to the participants and ascertaining that the outcome pathways would likely become concrete, considered, and plausible enough in the eyes of outsiders. The *channeling of action* metaphor seeks to underscore that the intermediate design considerations are not about openness per se but focus on ensuring that the participants had enough room to find direction—yet not to the extent of being diverted, distracted, or diffracted from the kind of outcomes they were promised—and an opportunity to deliver upon inviting them to participate.

Illustrating Intermediate Design with the Process of Designing MTPT

Let us next concretize these features through recounting the design process of the original mid-range toolset. Our initial aim was to simply adapt existing TA processes (Loorbach 2009; Roorda et al. 2012; Frantzeskaki et al. 2013) for the mid-range. Embarking on this, we soon noticed that more support would be beneficial. We first aimed at a relatively self-standing system of representation, but the design progressively incorporated more proceduralization in order to ascertain meaningful outcomes. At the same time, it is purposefully not a design game that can be "played" (cf. Torvinen 1999; Vaajakallio 2012) but rather a set of elements that the participants can appropriate in order to elaborate on and deliberate the change pathways and express their view of the pathway steps (Hyysalo, Marttila et al. 2019; Hyysalo, Perikangas et al. 2019). Getting there was however a long iterative process.

Among the first precursors to MTPT was a technology-specific forecasting workshop that was held in June 2016, with a focus on how different renewable energy technologies would develop in Finland by 2030. The workshop used an A1-sized board (see Fig. 4.7) that was divided into four different "scales" (household, village, area, and national) to aid deliberations on how the technologies might be materialized and what would be required to proliferate them effectively. Eventually the technology-specific boards were filled using color-coded sticky notes in a three-hour facilitated process performed by eight participant teams.

While the workshop got highly appreciative evaluations from the over 60 participants, we assessed afterward that the board would be too simplistic for forming the technology-independent pathways needed in the mid-range transition arena, where participants would need to:

- Formulate steps that would form a transition path for each transition goal
- Perceive the interrelationships between steps and the timing of needed actions
- Evaluate the realism of the suggested steps and the actions with which the essential steps can be supported
- Recognize the most critical steps in which societal choices must be made
- Consider the actions from the perspectives of different societal sectors

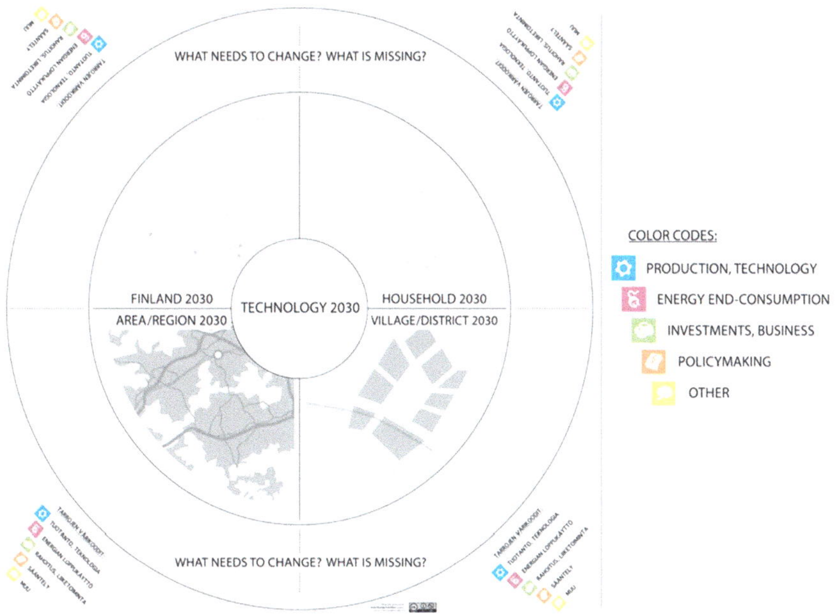

Fig. 4.7 The board that was used in the 2016 forecasting workshop [translated by the author, (Hyysalo, Perikangas et al. 2019)]

- Become better capacitated in understanding what potential actions people should be prepared for and what could be the means for achieving a transition.

The open questions and uncertainties were manifold at this stage, with multiple "goods" needing consideration, such as the quality of deliberation, the legitimacy of the arena process, the quality and integrity of the outcome pathways, the "insight value" that pathways would have outside the arena, participant ownership and commitment, the pace of work, and the number of pathways that could be pursued.

The design work was next continued through a master's level course held in late 2016 at Aalto University. Three student groups were presented with the earlier board and the concepts we had developed based on templates using the honeycomb structure on a circular board. The first student group elaborated on how the round board could include a sectorial inspection of different scales of the path steps. The paths were created by movable magnets to allow use of a vertical surface. The second group developed a highly ambitious yarn-based (and, by implication in a digital setting, digital connectors) cross-pathway formulation. The third group examined resilience and contingency factors related to pathways and pathway steps. They developed a notation system of four different elements that could structure deliberation over how to improve the resilience of the solutions through "what if" elements, adaptive measures, concrete actions, and the involvement of other actors. They further described a protocol for how these resilience deliberations could be organized in a hexagon form surrounding a particular pathway action (see Fig. 4.8).

From the first student group, the design team concluded that movable magnets would have several benefits compared to cardboard elements or sticky notes. Also, the way to agglomerate actions as additional elements around each pathway step was a useful idea, however it might be realized. As importantly, it became evident that structuring the work into four different scales as in the 2016 template (Fig. 4.7) would become too difficult to manage, as pathways could include actions on several scales, and a linking system between scales could hamper the fluidity of the work in fast-paced workshops. Finally, the potentially high amount and complexity of pathway steps in the pathways started to become evident and, at that, any circular notation almost necessarily ran out of the estate given the need to see text and human body dimensions. The level of complexity implied in the second group's encompassing multi-scale, multi-pathway construction board underscored that these more complex multi-pathway concepts just had to be scrapped. This would have been difficult to admit without students' designs that really pushed these ideas to their limits.

Consequently, design was shifted to a system for adequately representing just one pathway with a radically simplified notation system. After a few trials, we took a cue forward from the third student group's contingency and resilience board, which had been found to be too meticulous for its intended

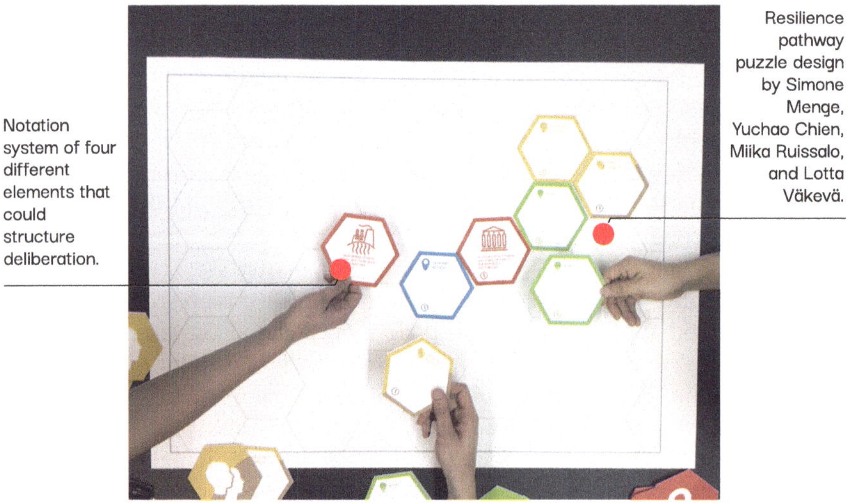

Fig. 4.8 The resilience and contingency consideration board by a student group that simplified the earlier hexagon-element-based pathway notation systems to open hexagon background and resilience elements. (Hyysalo, Perikangas et al. 2019)

task, as it would have required as much as 2–3 hours to deploy. Yet our team concluded that something like it could potentially work well as a basis for the pathway formation board itself. The separation of "pathway step" and "what actions are needed for the step?" elements was a clear solution, and this relatively simple design appeared versatile for the rather large uncertainties regarding how participants might work together and the now evident variety that the collaboratively produced pathways to different transition goals likely would have.

The first design sprint focused on the physical elements that might be needed for single pathway construction: what might be sufficient text visibility, element sizes, and the nature of the hexagon grid for a 4–6-person small group working on a pathway. Next, tradeoffs were experimented regarding the board and element sizes, resulting in the whole system expanding so it could include up to 20 steps and the associated "step actions" on a 150 cm × 240 cm board. In this design, the hexagon elements, not locations on the board, would indicate the information, scales, timing, and main actor for each step (see Figs. 4.1–4.5). We also realized that everything would need to be magnetic, including arrows and connectors as our playtesting made it evident that everything could be moved around quite a lot during the sessions. The team also had a suite of ideas that could be included, such as markers for pathway steps that converge with another path and ideas about how to mark (sub)pathways that contribute to several transition goals.

From here, the design strategy shifted into making iterative cycles of designing, prototyping, and testing, mixed with ideas from playtesting (Zimmerman 2003) and user dialogue (Buur and Bagger 1999). The designers collected ideas for a prototype and, when satisfied, built the prototype with which to run tests. This was followed by dialogue sessions with the test participants.

While the physical elements and facilitation procedures had been iterated in separation up to this point, only a realistic testing session with real content could give an idea about whether and how they might work together. Henceforth, the team manufactured all the elements from cardboard to create a mock-up, wrote a facilitator guide (see Fig. 4.9), and selected two transition goals from the Finnish Energy and Climate strategy likely to feature in the arena.

This mock-up test verified that the design of elements and the board just needed minor tweaks, but it also raised new issues. First, it became evident that the pathways could require as much as 20 steps, each with 5–10 enabling actions, and the pathways could thus become more complex than expected, for instance, by containing several converging and diverging sub-paths. Also, the step interrelations could become more varied than what could be marked with just plain arrows; hence, arrows with a writing surface were designed. Also, the order by which the pathways would be constructed on the board was subject to ambiguity, potentially calling for a more structured process rather than leaving the progression to the participants and the facilitator as originally

Fig. 4.9 The prototype for pilot 1 was prepared by printing the MTPT elements on cardboard. (Hyysalo, Perikangas et al. 2019)

envisioned, yet any guidance would have needed to retain enough openness to make the work empowering, effective, and fun.

The mock-up test revealed that the pathway construction was slower yet more complex than anticipated. The TA team had assumed that the steps would have been relatively well-known to all, and just by and large deployed on the board with some accompanying discussion, and that the real new insights would follow from pathway interrelationships. The industry expert on the design team, however, pointed to many areas that were likely to be shrouded with uncertainty, thus taking time from the participants, however expert they may be in their own specialist areas. This was further related to a revelation that there would be a considerable need for fact-checking during pathway construction and thus underscored a need for an information package to be given beforehand. Finally, it became evident that considerations about the feasibility of the suggested actions might require some type of focalization for each pathway. To aid this, future cast personas were created, encouraged by the Melbourne resilient futures project (Gaziulusoy and Ryan 2017b).

Most of these test insights were not related to the physical notation system per se but to how the process and procedures were to take place. It became clear that the pathway board and its elements paired with a facilitator would leave the process too vague given that the participants were busy and high-profile people, such as members of parliament, city mayors, CEOs, and civil society actors. The ensuing design actions thus became primarily "social" and "procedural" rather than changes to physical elements: dividing the work into several phases, creating facilitator and participant guides, and introducing background memos.

After these redesigns, the second testing ran with four experts on the energy systems transition working in the same research consortium. The path creation was facilitated according to the planned run-through, performed on a horizontal cardboard pathway board, and featured cardboard step elements, vision persona, a goal description, a background memo, and a participation guide. Three notetakers tried different recording and note-taking options. This beta-level testing confirmed that the toolset by and large worked, but most discussion now revolved around what exactly the pathway construction was to achieve. It was now evident that the elaboration of a pathway could take up to 2–3 hours, over three times as long as first anticipated. The question was whether to aim for a few well-elaborated paths or to aim for a meaningful share of potentially 30 different paths required to depict an entire Finnish energy transition.

The questions about alternatives in the pathways further underscored how to duly address contingencies and alternative pathways to transition goals. The eventual solution was to extend the pathway creation process from two to four phases, proceeding to mark, in phase 3, how much uncertainty there was in timing or actualizing an individual step with the help of uncertainties elaborated in the background materials. Phase 4 would consist of creating alternative paths.

These iterations to MTPT were worked into the participant and facilitator guide, with added emphasis on how to deal with uneven levels of expertise among the participants—without, for example, limiting experts' "airtime," all participants needed to be guaranteed enough space to voice their views and to deliberate over the steps and actions in the pathway.

The test also revealed that notetakers could not record the rapid conversation, which meandered between deliberation and discussing what to write on each pathway element. The documentation thus shifted to having video shoots after each phase of the process, as this would give participants a clearer voice in explaining the logic of the envisioned pathway.

The last testing session was a month before the first pathway formation workshop with magnetic boards and elements and facilitator and participant guides ready. Here, the focus was on the resilience and alternative pathways' construction phases, which were hotly debated regarding their credibility and validity.

In retrospect, the iterative prototyping and testing of MTPT revolved around balancing multiple competing "goods"—such as elapsed time, the quality of outcome pathways, the amount and kind of interactions and deliberations between participants, and the freedom they would have in constructing the paths—in how the physical elements were to be put into motion through procedures and facilitation and through minor modifications to the physical elements. Participants' ability to formulate and deliberate pathways with the toolset was evident in both their actions and in their later evaluation. Many explicitly underscored how the toolset helped to see the systemic nature of the needed changes, to perceive interrelations between pathway steps, and to share expertise across the small groups (Hyysalo, Marttila et al. 2019; Hyysalo, Perikangas et al. 2019). The identification of the most critical steps also worked well, forming clear spearheads for action amidst the 100 + actions identified in the final 200-page report (Hyysalo et al. 2018).

Pathways remained mostly within the parameters that the design team had tested and designed for: fewer than 25 steps, several parallel sub-pathways that came to affect each other at some point, fewer than seven prioritized steps in which actions were specified, and 10–40 pathway actions. This variation in pathways and the robustness in accommodating different types of systemic changes were perhaps the clearest evidence that the intermediate design had succeeded. It capacitated the participants to form ambitious yet viable pathways of which they took ownership as a group, which reflected the social changes in the examined domain rather than a pre-conceived notion of change into the system (Hyysalo, Marttila et al. 2019; Hyysalo, Perikangas et al. 2019). Yet, the time use remained an issue, being 3–8 hours per pathway in the "real" mid-range arenas (Table 4.1).

Chapter Conclusions: Design Participation for Transition and Transformation Governance

Design participation is increasingly used in catalyzing society-wide changes in futuring and transition process-related deliberations. These processes underscore the role of intermediate designs that help participants to reach meaningful outcomes in the face of high complexity and divergent participant perspectives and where the key objective is to give participants freedom to deliberate, express, create, and take ownership of the process and its results (rather than resulting in a finished design or the designers taking center stage in the realization). Intermediate design in such processes, and potentially beyond, is aptly characterized as channeling participants' action.

For practitioners, our analysis underscores how creating working intermediate design requires an iterative process and refraining from unduly designing in users or their participation, while ensuring that the design elements and process will create the needed quality and quantity of outcomes. This requires careful balancing of participation ideals and mundane practicalities; reducing the complexity of the intermediate design within the minimal complexity required by the target system; adequate target domain knowledge; distribution of agency between participants and facilitators; catering for multiple potentially conflicting goods and goals; and blending the design of physical elements, procedures, interaction patterns, facilitation measures, and documentation requirements. Failing any one of these facets is likely to result in a collapse in the use of the intermediate design or in unsatisfactory results. This new breed of design participation thus presents an intriguing yet challenging area for design practitioners and researchers.

Table 4.1 Phases of intermediate design in the MTPT process

Stage of iteration	*Participants involved*	*Development of the tool*	*The focus of the intermediate design process*
Initial stage	The TA teams with student teams	• Basic ideas for the board and play materials are introduced • Initial ideas for path creation activity	• Iterations of existing TM methods • Materials and their contextual and physical qualities
Second stage (pilot 1)	The TA teams and some colleagues as test participants	• The board and material mockups are introduced • Initial "rules" for playing are introduced	• The complexity of pathways, slowness of working • The need for a more structured process of working and pre-structured information
Pilot 2	The TA team with SET project experts as participants	• A play guide is introduced; four phases of work are introduced • Structured documentation takes place	• What is to be achieved by TA? • Addressing contingencies and alternative pathways • Ensuring space to voice all views
Final stage (pilot 3)	The TA team with SET project experts as participants	• The process and play guide are revised with minor iterations	• Final iteration of the tool • Managing the facilitation load

Regarding designing for transitions, our work illustrates that there is plenty of important work that designers and design researchers can pursue to enhance the main avenues of transition governance that have been set in motion by social scientists. While transition governance has a considerable multidisciplinary community and a history of analyzing and fostering long-term systemic change (and it may well be illusory for design researchers to seek to ideate replacements for these models), the means used to facilitate these complex processes benefit from targeted design efforts (Cf. Gaziulusoy and Ryan 2017a, b; Lähteenoja et al. 2023). Particularly, drawing on the design traditions of creating mediating representational artifacts, toolsets, and procedures for multi-stakeholder participation can improve participants' time use, resulting in higher-quality outputs and giving participants more agency and more direct agency in the transition governance processes (Cooney et al. 2018; Eriksen et al. 2014; Hyysalo, Marttila et al. 2019; Hyysalo, Perikangas et al. 2019; Agid 2018).

Sustainability transitions affect wide constituencies of society and, as Voß et al. (2009) point out, this calls for wide societal engagement. In other words, it calls for various forms of designing for wide societal participation, governance, and social change, which is an area of rising design research interest (e.g., Agid 2018; Hyysalo and Hyysalo 2018; Julier and Kimbell 2019). Our experience underscores that designing for governance greatly benefits from, if not requires, multidisciplinary competencies. The Finnish mid-range arena processes have been pursued in large organizer teams that include experts in political science, policy analysis, and sociotechnical change, who have provided crucial insights to the design and deployment of mid-range arenas and the toolset (Hyysalo, Marttila et al. 2019; Hyysalo, Lukkarinen et al. 2019; Lähteenoja et al. 2022; Lukkarinen et al. 2023; Lähteenoja et al. 2023; Marttila et al. 2023; Lukkarinen et al. forthcoming). The in-depth understanding of policy processes, actor remits, and persistent and current challenges in different governance institutions has been vital for our design, as was their long experience of interacting with relevant civil servants, politicians, businesspeople, and NGOs. This domain knowledge was used to anticipate the issues that needed particular attention, tensions between participants, tuning facilitators in the workshops and estimating participants' available time allotments, attainable goals, and so on. Storming in with a team comprized of only designers would have been rather less likely to succeed.

In all, sector-wide sustainability transitions present new contexts and challenges for design participation in terms of scope, longevity, ownership, and types of outcomes sought. As only handfuls of concrete design projects concerning transformative change can yet be found, it is time to roll up our sleeves, as there is no shortage of design contributions needed to be made in support of sustainability transitions.

CHAPTER 5

Diversifying Participation with Mixes and Ecologies of Knowledge

Introduction: The Idea of and Rationale for Participation Mixes

In this chapter,[1] we expand our discussion of design participation from particular approaches to considerations of how to diversify and mix different participation formats. Mixes of participation formats can be useful in contexts in which any one single approach to design participation is likely to remain inadequate. We argue that such situations are quite common outside academic settings. We introduce the idea of participation mixes with the help of a large-scale participatory effort in the planning of Helsinki Central Library Oodi. We then take a further step and introduce the idea of ecologies of knowledge mapping as a way to further analyze and tailor the participation mixes in an organization.

It is not always, even in principle, possible to identify and attract the "right" participants or even the "right blend" of participants to a design participation. Potential participants may simply have too widely different motivations, practical possibilities, and limitations to participate, as well as widely differing background knowledge and design competencies to be incorporated in any

[1] Earlier versions on the analyses going into this chapter have been published in three articles that have been thoroughly rewritten for this book: Hyysalo, Virve & Hyysalo, Sampsa. (2018) Mundane and strategic work in collaborative design. *Design Issues* 34(3), 42–58; and Hyysalo, Sampsa, Hyysalo, Virve and Hakkarainen, Louna. (2019). The work of democratized design in setting up a hosted citizen-designer community. *International Journal of Design* 13(1), 69–82; and Mäkinen, Samuli, Hyysalo, Sampsa & Johnson, Mikael. (2018) Ecologies of user knowledge: linking user insight in organisations to specific projects. *Technology Analysis & Strategic Management* 31(3), 340–355.

S. Hyysalo, *Design Participation*,
https://doi.org/10.1007/978-3-031-95437-5_5

one "format" of participation (Jonhson et al. 2014b; Hyysalo, Jensen et al. 2016; Chilvers and Kearnes 2016). The differences among the people relevant for design may simply call for very different arrangements for participation.

And, just as well, there may be important differences in the ends to which different types of design participation are sought. Shoehorning all the potential participants and reasons for design participation into any one participatory format will do injustice to some as well as lose opportunities however cleverly and painstakingly it is designed (Johnson 2013; Johnson et al. 2014; Mozaffar 2016).

This was also one of the core reasons why in the introduction we distanced ourselves from any one normative idea of what public participation ought to be and how it should be realized. We rather emphasized that meaningful participation in different contexts, for different issues, and with different people is always a situated practical (and often anxious) achievement across several competing "goods" for what would constitute high-quality and impactful participation (Friedman et al. 2012; Chilvers and Kearnes 2018; Hyysalo, Jensen et al. 2016). In Marres's (2013) terms, the different formats of participation affect how *relevant* the engagement with design participation becomes for different people, thus offering an "equal" participation format, as in offering any one format equally to all, neglects the fact that it then necessarily becomes unequally relevant for different potentially implicated people.

One avenue toward resolving this quandary is through building "participation mixes," which are built to offer complementary ways of design participation to enable meaningful participation across differences present among potential participants; differences in these people's preferred ways of participating; diversity in the design and knowledge needs present in the design project(s); differences in potential participants' design ability and willingness to engage in design; and often (the finite) resources available for organizing participation.

Against these characteristics, a successful participation mix:

- Has identified the matters and issues involved and how they may relate to different people
- Has identified the potentially affected and implicated people who could somehow be attracted to participate
- Has identified the time–spaces for meaningful participation and for what and with whom the design participation then could be sought
- Has created sufficient (and sufficiently diverse) channels and formats of participation that in practice make participation possible and desirable for those wanting to participate (or wanted to participate)
- Has worked out how different participations can make a difference and complement (or balance or contest) each other and has built a logic across them
- Can effectively further the design endeavor with the participation mix and implement the insights and requirements duly.

As such, the idea of participation mixes is not new, but it has remained less elaborated than it would have deserved. On one hand, medium and large participatory and human-centered design projects have always involved conducting complementary user research and user-involvement activities either in sequence or in parallel to address different users and knowledge and design needs of the projects (e.g., Bødger et al. 2004; Dalsgaard et al. 2012; Hyysalo, Repo et al. 2016). On the other hand, large private companies have used their user groups to similar effects, segmenting them according to voluntary participants' abilities and interests in, for example, future trend scanning, concept development, usability and UX testing, end-user innovation, and prioritization of development requests (Prandelli et al. 2008; Sawnhey et al., 2005; Antorini et al., 2012; Johnson, 2013; Mozaffar 2016; Pollock et al. 2016; Pollock and Hyysalo 2014). In both types of cases, reporting the participation mixes has remained in the mode of what was done and elaborating on the benefits gained for the design project, the patron organization, and sometimes also the participants. What remained missing was reflections of the rationales used for building up the mix in one way rather than others and the effects of these choices.

Furthermore, as we discuss in the latter part of this chapter, in most such discussions, the resulting complexity in how different participations and ways of knowing how the citizen/user/customer insights relate to organizations and institutions has been treated superfluously. In other words, what we call resulting "ecologies of user knowledge" have not been paid due attention as starting points for and long-term achievements resulting from design participation. This is particularly important in cases where an organization wishes to deepen its engagement with design participation. Mäkinen et al. (2018) elaborate on how participation mixes and ecologies of user knowledge concretely help in the following activities that organizations engaged in design participation commonly pursue:

- Assessing if a project may be missing insights from already gathered information about customers
- Assessing if a project could benefit from the already existing repertoires that customers/users are related to in the organization
- Identifying systematic gaps in design participation and/or user research at the project level
- Identifying the organizational structures and dynamics that lead to dysfunctional patterns of participation and/or user research method choices
- Identifying the dynamics that (may) hinder the use of potentially more effective design participation engagements and formats
- Assessing an organization's design participation mix and making comparisons across projects and organizations.

To elaborate on these issues, we shall first examine building the participation mix for Helsinki Central Library Oodi and, after this, move on to elaborate on the import of ecologies of user knowledge with two cases at the National Finnish Broadcasting Company.

Oodi Design Participation Mix

Public organizations have much to gain from involving citizens in their design and development activities. Possibilities to participate in shaping one's city or country add to democratic opportunities for citizens. Involving citizens in the development of public services is increasingly seen as a way to develop more adequate and equitable services and presents a voluntary resource for public authorities in the design and delivery of public services. Consequently, the use of service design and codesign in the public sector have become more widespread during the last two decades (Bovaird 2007; Junginger 2008; Bovaird and Loeffer 2012; Ansell and Torfing 2021).

In the early 2010s, the City of Helsinki committed to participation, open innovation, and citizen involvement at the strategy level to increase the possibilities for the residents to influence the making of a human-centric city (Helsinki City Strategy Programme 2013–2016). This new strategy condensed a marked shift in the 40,000 employee city organization that had often been seen by citizens as an inflexible behemoth regarding, for instance, grassroots citizen action and new urban culture. The aim was to transform the city organization to become more flexible to citizen views, particularly in its planning processes, and to bestow more power in the hands of the elected city board over the planners in its bureaus. The increasing use of service design and design participation has resonated with this strategy and has been taken as a key means to further it (V. Hyysalo 2022; Pirinen et al. 2022; Hyysalo et al. 2023). The participatory planning of the flagship Central Library Oodi has been one of the most ambitious and influential projects shaping this development.

Libraries continue to be in transformation across the globe. Instead of being just repositories of and access points for books and other cultural productions, they are becoming community hubs and important third places (Dalsgaard 2012; V. Hyysalo 2022). The functions include serving as innovative and inspiring alternative working spaces, acting as enjoyable community centers, providing opportunities for various forms of cultural production, and serving as platforms for people's own initiatives. This entails two new ways for libraries to engage in design participation in relation to their customers: design participation in the design, planning, and deployment of new facilities (such as codesigning Oodi) and design participation that these new facilities enable among and between citizens and the city administration (such as those enabled by Oodi) (V. Hyysalo 2022).

In keeping with this renewal, participatory planning was part of Helsinki Library Services' preplanning and planning process for Oodi from 2012 to

2015, leading to the inauguration in 2018. From the beginning to the end, it had dual objectives: in addition to contributing to the planning process, citizen and stakeholder participation offered a way to conduct the communications and marketing of the yet formally undecided project in an interactive and dialogic manner.[2] The preplanning phase featured a participation mix of 10 different complementary forms of design participation (Table 5.1). These activities used different channels to reach citizens and had a range of complementary aims and depths of participation. Consequently, they produced different outcomes and materials to support the planning process, ranging from gaining 2600 "library dreams" to setting up a volunteer citizen-designer community (Hyysalo et al. 2014; Hyysalo and Hyysalo 2018; Hyysalo, Hyysalo et al. 2019; V. Hyysalo 2022). Some of the initiatives were trailblazing in Finland, such as opening architectural competition entries to public view and voting to inform the jury and large-scale participatory budgeting and which were both first-of-a-kind in Finland and have since been actively used in the country.

Assessment and Planning of Participation Mixes for Oodi

The design participation activities began with assessment of the public and communities relevant for different service offerings at Oodi. Helsinki Library Services had a good understanding of their current customers and their use of libraries, and thus, much was already known about the usage, users, and design solutions for different service offerings by the library planners. But Oodi was also to transform what libraries would offer, while new generations of users would continue to have new preferences for digital, learning, and making facilities every few years.

The participation planner of the Oodi project (Virve Hyysalo) and the author (who consulted in the planning and executing the design participation actions) started off in 2012 with a mapping exercise to identify which service offerings and spaces indicated by the preplanning document were already well covered in terms of knowledge of how to design them and where the knowledge shortcomings, uncertainties, and unresolved issues were. The mapping focused on which publics and communities were already well represented in the library's ongoing participation efforts and which ones less so, particularly with Oodi's new service profile in mind. The results of this relevance and framing work underscored that the design participation activities had diverse audiences and particularly that there was a need to address the general public in a relatively undifferentiating manner. At the same time,

[2] Helsinki Library Services was in charge of the content and space reservations in which the participation activities focused, while the city planning office was in charge of the allotment and building specifics, culminating in an open architecture competition won by ALA architects in 2014.

Table 5.1 Collaborative design activities in the Oodi preplanning stage

Codesign activity	*Aim*	*Channels and tools*	*Target groups*	*Timespan*	*Level of participation*
Tree of Dreams, idea crowdsourcing, and citizen discussion	• Raising awareness and rebranding the library • Collecting unexpected ideas • Empowering citizens	• Digital platform • Outdoor advertizement campaign • Urban events	• General public • Challenging target groups • Non-users, potential users	March/12-Dec/13	3
Events in different branch libraries in Helsinki area, Signal us! campaign, Tree of Dreams campaign	• Raising awareness of CeLib • Introducing more customer-centric ways of working • Empowering citizens	• Events, campaigns, workshops	• Library users • Loyal customers and so called "library lovers"	Sep/12–Dec/13	3
Urban experientiality of future library	• Rebranding the library • Creating and testing new services • Building closer customer relationship • Building community feeling	• Urban events (i.e., World Design Capital Helsinki 2012 year) • Building pop-up labs and environments	• Challenging target groups • Non-users, potential users • New audiences	Sep/12–May/13	2–3
Invitational workshops	• Creating solutions to demanding planning questions • To reach target groups identified as crucial for planning	• Workshops, events • Creating small communities around planning questions	• Challenging target groups (e.g. teens, families, and children, maker spaces, immigrants)	Feb/12–May/14	3–4

Codesign activity	*Aim*	*Channels and tools*	*Target groups*	*Timespan*	*Level of participation*
Stakeholder network	• Creating partnerships, collaboration, and communities	• Workshops, events for meeting, mingling, and ideating together	• Stakeholders and partners to CeLib	May/13–March/15	3–4
Open architecture competition (in two phases)	• Opening up the competition entries (over 500) for voting and commenting	• Exhibitions • Interactive displays around city • Web participation	• General public • Architecture and design enthusiasts	March/12–April/12 May/13–June/13	3
Participatory budgeting and pilots in new libraries (with partners Emobit and Sitra), total amount €100,000	• Empowering citizens • Creating pilot projects together • Sharing decision-making power	• Workshops and online participation • Transparent decision-making rule	• General public • Citizen activists	Oct/12–May/14	4–5
Bicycling pop-up mini-libraries (with six different themes)	• To bring library activities to people in the streets of Helsinki • Celebrating the closing of the architectural competition	• Urban event and pop-up cargo bikes all over the city	• General public • Non-users, potential users	June/13	2

(continued)

Table 5.1 (continued)

Codesign activity	*Aim*	*Channels and tools*	*Target groups*	*Timespan*	*Level of participation*
Idea nugget, idea competition both internally and externally (for staff and citizens)	• Collecting unexpected ideas • Creating new service solutions	• Marketing campaign • Sparring ideas	• General public and staff • Inventors and developers	Sep/13–April/14	3–4
Friends of the Central Library, citizen designer community	• Informing planners, empowering citizens • Testing ideas, creating new service solutions • Introducing more customer-centric ways of working	• Working together over an intense three-month period • Workshops and online participation	• General public	Oct/14–Dec/15	4–5

Note The depth of participation scale used is based on the IAP2 Spectrum for Public Participation by International Association of Participation https://www.iap2.org/page/pillars

future digital services, peer-to-peer learning, maker spaces, gaming environments, future work spaces, and cultural production spaces would benefit from targeted participation measures to clarify what these areas of the future library would consist of.

As Table 5.1 documents, different design participation activities had different aims as well as different channels through which to reach potential publics. The channels to reach people were paired with different participation formats, methods, and tools to realize the participation. Following from these, the quality of participation varied as well—whether it was limited to expressing ideas and consulting the design team or reaching out to joint design solutions and decision-making by the public as, for instance, in some of the invitational workshops and in the participatory budgeting.

The eventually realized mix featured both activities planned from the outset as well as seizing some emerging opportunities, particularly those opened by the World Design Capital Helsinki 2012 (WDC) public events and urban democracy initiatives. Joining these typically high publicity events was further encouraged by the dual strategy of participatory planning and participatory marketing of the Oodi construction project. Yet none of the participation activities was improvised or purely reactive, and they were initiated only once they matched the rationales in our schematic as to what design knowledge and public engagement was to be sought or was still missing.

It is also worth recapitulating that, in such a real-life project, the estimations on costs and gained results always feature importantly. In this regard, while planning the participatory activities, important heuristic considerations were, on one hand, the simple "contact price per customer" typically used in marketing and, on the other hand, a more design-oriented "engagement price-per-design contribution" as proxies for assessing what might be the value for the progression of Oodi's planning. Importantly, these two heuristics often ran contrary to each other, as meaningful design contributions tended to require deeper and more time-consuming engagements.

The participation activities began in the spring of 2012 in two parallel avenues at two extremes of design participation. On one hand, a series of focused workshops was launched to target the most pressing design needs of Oodi planning such as the maker spaces, peer-learning facilities, and digital service offerings. On the other hand, the design participation reached out to the general public via a multi-channel endeavor. Let us examine these and their yield in more detail, starting from activities aimed at the general public.[3]

[3] This analysis is based on the documentation and assessment discussions we conducted throughout the design participation processes for Oodi. We ran informal assessment discussions after each participation event based on the authors' notes, and a more formal "post-mortem" discussion was held once the dust had cleared after each activity. The compiled notes were cross-compared in November 2014 and annotated in 2015 after the user-designer community "Friends of Central Library" in January 2015. The analysis by Friends of Central Library was complemented by notes by an independent observer (Louna Hakkarainen) from the second author's research group (see Hyysalo et al. 2019) and a

Open Ideas Harvesting: Digital Platform and Advertising Campaign

Participatory planning activities started with an open call for ideas about what the future library should be and what could be in it. To avoid people being locked into incrementalism and current library services that commonly feature in idea campaigns, the campaign asked for "library dreams" from citizens rather than just ideas. To raise awareness of the Central Library project, the dream campaign was initiated with a poster campaign in public advertising spaces and libraries where known Finnish cultural figures shared their library dreams and which included an invitation to leave one's own library dream in any library or on the digital "tree of dreams" on the Central Library website. Citizens responded by posting 2300 library dreams during the years 2012–2013.

In a retrospective assessment carried out after the campaign, the campaign and web participation had several upsides. It paired marketing outreach well with initiation of participatory planning. It was a new kind of opening to invite citizens to influence planning by spelling their dreams and ideas and making all dreams openly visible to all through the digital platform. This "tree of dreams" was found visually appealing and created an easy channel for citizens to participate. The decision to have an advertising campaign in downtown Helsinki was also good for portraying the future Central Library contents and activities, encouraging participation, and, together with the platform and tree, emphasizing how the envisioned library was to become a combination of novel digital services and production spaces and more traditional library spaces. The digital recording made it possible to accumulate contact lists of interested citizens with names and emails and, in principle, enabled digital handling of materials.

The web participation did not, per se, require new competences from Library Services beyond the work of its participation planner, but in practice this turned out not to be the case. Here we come to the downsides. The open call for dreams did not target any of the focus areas as to where most new insight for planning was needed from the citizens. Few citizens used the tagging option available, and those who did used a wide variety of terms, in all resulting in an undifferentiated mass of ideas and dreams. The content analysis of dreams was hoped to be gained from a partner company, but the results remained too superficial and cursory to the librarians' and designers' eyes. The web participation created a one-way information transfer, with no mechanism for refining or gaining background information to the shortly expressed ideas. These shortcomings hampered the yield of the web campaign but became partially mitigated by two adjoining actions that Library Services took, albeit

separately researched interview-based thesis on a participant's experiences in the user innovation community (Hyödynmaa 2016). We also ran interviews with the involved planners during the process, and the primary notes and documents of the Central Library planning group were used for double-checking the details.

inducing considerable added work in refining and making sense of the gained content.

Open Idea Harvesting: Urban Events in World Design Capital Helsinki 2012

The "dream campaign" was complemented by Library Services' participation in a series of urban events that were held during the WDC. The library planners and librarians held a Central Library pop-up spot in conjunction with a range of events, including local street art festivals, the WDC Pavilion event series, a pop-up sauna, yarn bombing and Maker Faire events, and running a day with pop-up bicycle libraries and a party at the allotment site, all of which connected to the contents of the future library. In all these events, library dreams were collected from interested participants. The tree of dreams was displayed as a physical embodiment either on a real tree or using Finnish design icon Eero Aarnio's space divider tree (Fig. 5.1) and cards in which to write and display the ideas. The cards and campaign posters were also visible in over 30 branch libraries.

The physical encounters with people allowed for gaining knowledge of peoples' backgrounds and contexts of their dreams, which the web participation did not provide. In the words of a librarian, these interactions resulted in "higher-quality dreams," as they could encourage the citizens to write more of this context on the cards. The events created contacts for library planners to new groups of people that did not visit libraries that often, such as urban activists, some youth groups, open data enthusiasts, and designers. They also led to interactions with opinion leaders such as local and national politicians and celebrities. The event participation created a possibility to refresh the public image of Library Services among these people. The new library was included in favorable contexts and linked to the future orientation prevalent in the WDC events in a lively fashion.

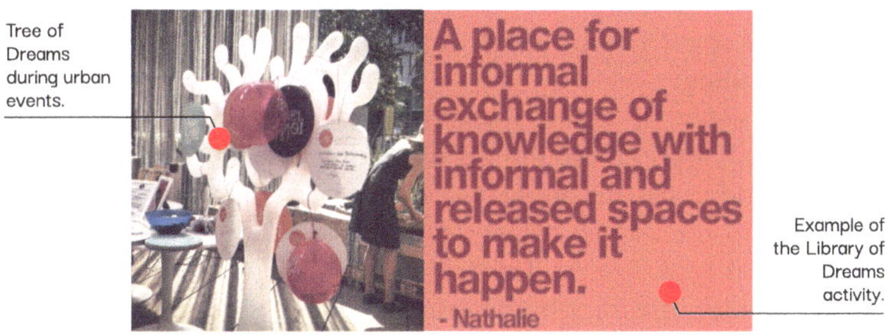

Fig. 5.1 The physical tree of dreams (an Eero Aarnio design) and an example of a library dream (Photos: Virve Hyysalo)

The event participation was heavy on library staff, as most events were held on evenings and weekends. The staff felt that the events were difficult to anticipate or plan properly, which added to the needs to improvise and adjust. Event participation was also relatively resource-intensive as contents needed some customizing for different events. The dreams written on paper induced the burden of inserting them into the database afterward and the dreams and ideas were still just ideas—not ready solutions or concepts. The library planners thus concluded that the 400 "higher-quality dreams" came with a considerably higher "contact price per customer" than the ones collected online.

For the library staff, events expanded the diversity of people with whom they had contact in library renewal efforts, with many promising contacts, even though the planners were ill-equipped to record the contacts systematically for further use. The part of the campaign that ran in the neighborhood libraries added to the events, but on the whole, its yield remained lower than expected—higher yield might have required that the branch libraries would have really actively stopped and drawn visitors to the campaign rather than only displaying it and having the idea collection take place alongside all the other things going on in their premises.

"The Dream Job": The Work Needed to Analyze the Library Dreams

As noted, the purchased analysis of library dreams remained too cursory for feeding in further design. To be able to utilize the materials, the participation planner contacted her colleagues from library staff to participate in "dream job," a two-day sprint to sort the 2700 library dreams into themes and to generate content descriptions of the most salient themes. This thematic analysis across the dream data raised to the front eight themes, each of which were then turned into descriptions that included a general explanation of the theme and description of its subcategories, a quantified summary, and illustration of the contents by displaying individual good ideas. The theme descriptions were deemed illustrative and well grounded. The themes and ideas clustered under them offered a backbone for developing a series of pilot projects for branch libraries in 2014 and feeding further into participatory budgeting along the main library planning.

On the downside, the theme formation was merely the first cut through the wealth of data, and the full utilization of dream campaign contents would have required additional means, which were practically unattainable. At the same time, even as the themes were effectively formed, the finalization of each theme description took several days from the participation planner. In terms of staff participation, dream job was instrumental in spreading awareness and enthusiasm in more direct engagement with citizens among the library staff but again was limited to those who were in contact with participation activities.

Invitational Participatory Workshops and Events

Open call participation was complemented with invitational workshops, as it was recognized that the future library had several areas where planners needed complementary in-depth understanding and enhanced dialogue with affected peoples, stakeholders, and experts. As noted, we spent several days assessing the existing knowledge that library planners already had regarding the different users of spaces and activities envisioned in the preliminary project plan. As a result, about 20 invitational events for 13–50 people were organized in the following areas: child care and child pedagogues; literacy, book, and print actors; tourism; multiculturalism; physically neighboring institutions to the Central Library; youth and youth services; families with children; peer learning and doing; digital media; and lead users in digital–physical making activities.

These events provided a way to reach the most crucial actor groups regarding the planning challenges. They provided an opportunity for focused engagements with the participants and produced insights that were directly relevant for the identified knowledge gaps in Oodi planning. The knowledge that was gained contained many solution ideas and important background information. The network events allowed for multi-directional exchanges between the invited participants and the planners, and the content specialists from the library gained new knowledge and connections. From a marketing perspective, the events created buy-in among these actor groups and allowed the library to elaborate on what the future of Library Services could be comprised of (Fig. 5.2).

While the participation events were seen as highly useful, preparing for them required a considerable amount of work in identifying the relevant people to

Fig. 5.2 Invitational workshops typically had 15–50 participants. The picture is from the maker space workshop in 2012 (Photo: Virve Hyysalo, reprinted with permission)

contact and then making the contacts. To gain a good yield in the short time of a half to full day that was attainable by the expert participants required days of preparing and rehearsing. This was preceded by further testing and iterations before well-suited formats were found regarding what was feasible to accomplish in the time-slot as well as what the key priorities were in hosting the events and in terms of their outcomes.

The expert and stakeholder workshops provided an excess of ideas and solutions in terms of what Library Services could take forward. As with the open call for library dreams, most ideas and information could not be effectively stored or acted upon beyond the most clearly relevant ones. At the same time, the post-mortem evaluation concluded that if invitational workshops had been the only engagement form with users, some amount of "out-of-the-box" input regarding the future library that was gained in open idea gathering would have not been received. Moreover, the participant groups invited to events and workshops, albeit many, were still just those related to the Library projects' knowledge needs for particular service areas and actor groups.

Participatory Budgeting of Development Funds

Helsinki City Library was the first organization to run large-scale participatory budgeting in Finland (with the help of private-sector partners and the Finnish Innovation Fund) in late 2012. Citizens were given the opportunity to collectively decide how most of the library's annual development project money for 2013 should be spent, which was altogether €100,000. The Central Library project created eight pilot concepts based on "the dream job". The residents were also given the opportunity to suggest their own projects or modify and develop further the eight suggestions. In the last phase, they voted on which pilots should be realized based on pre-budgeted cost structures. Citizens could participate either on the web or in one of three workshops that were held for the budgeting.

Partners created together a simple method and frame, and with the help of it, citizens could take part in the process. The frame included a method for decision-making and a toolkit for workshops as well as visualization of the budget to help the citizens interpret the data. After the web participation and three different workshops, the results were analyzed and announced to the public. The four pilot projects chosen to be implemented were the Urban Workshop maker space concept, Storybook Birthday Parties for families and children, Space for relaxation and concentration, and Lost and Found literature event series, making for a total budget of €108,000. To stay within the framework of €100,000, the participants decided that Space for relaxation would be implemented in a lighter and more inexpensive way.

The participatory budgeting gained a lot of positive media attention, as it moved beyond mere public consultation. As citizens actually decided on the budget allocation, the concrete decision-making power was distributed to the public, which was hailed as positive by the participants, library staff, and media alike. Participatory budgeting also gave an opportunity to refine, develop further, and concretize the dream data collected from the citizens into eight pilot project concepts. The pilot projects were a concrete promise from the library to realize some of the dreams as experiments quickly in a one-year schedule and also to learn from the experiments so that the concepts could be developed further. The majority of the realized projects were a success; for example, the Storybook Birthday Parties were booked out in just four hours and the Kaupunkiverstas maker space was one of the library's most loved services of the year.

The framework for co-working with citizens, the decision rules, and workshop methods proved to work out well, and there was enough flexibility in the process so that citizens could both suggest their own project models and say aloud their wishes and feedback. The participants ranged from stay-at-home-parents to coders, from urban activists to literature enthusiasts. The atmosphere in the participatory budgeting workshops was not all fun and games and at times, ardent discussions were needed to bring consensus.

In our post-mortem analysis we considered that the number of participants in the workshops could have been increased—only 60 citizens decided which pilot projects should be launched. Some attendees had also mentioned that the name "participatory budgeting" drove some people away because it sounded like the minimum requirement to join in was to be able to justify your opinions like a professional politician. On the library side, the timeframe of the project was very tight—only four months from idea to realization with the first-ever run of the economic data visualization, web participation, and marketing. The schedule affected the library's team—had members come from different branches and levels in the library organization, the results and their implementation would have been even better. Now there was a gap between the team who organized the participatory budgeting and the teams who were supposed to implement the pilot projects, and having some of the implementation team members' present in the budgeting events could have helped in transferring knowledge and increasing commitment to the pilots. The longevity of some the pilots suffered for this reason. For example, the Storybook Birthday Party was utterly loved by customers but straining on the staff, which eventually prohibited making the service permanent. In hindsight, it would have required collaboration with the city's educational and child care division or with university child pedagogue students to help engage and practically manage the groups of children.

Oodi's participatory budgeting was important as a pilot in participative democracy in Finland as the city proceeded to make it a permanent annual feature for the development of its four major urban areas, with a refined cross-city division procedure and €7 million allocated budget. This in turn has inspired other Finnish cities to engage with it.

Signal Us! Campaign

Among the more lighthearted participatory activities pursued by Helsinki Library was the joint campaign Signal Us! in all the branch libraries, where customers could give feedback by placing Post-its directly onto all items in the library spaces. The event aimed at spreading design engagement with citizens beyond the Central Library project and to complement the dream campaign in local libraries. Its format was an adaptation of a full-scale prototyping technique used earlier as part of a lead-user workshop with maker activities (see Chapter 2; Hyysalo et al. 2014; V. Hyysalo 2021, 2022), the point being that the reference to suggestions would make more sense when the comment was directly placed onto the item it concerned.

In the post-mortem we assessed that Signal Us! proved an easy and fun direct opinion/influence channel for customers. Generating and posting ideas directly in a library space was reported to have provided stimuli for development ideas from customers. For staff it offered clearly targeted ideas—in other words, the “aboutness” of development ideas was very good. The feedback gathering was also relatively practical to run amidst everyday work. It further engaged branch library staff into intensive customer (development) interactions and into sorting and thinking about development ideas, both of which were new tasks to many.

A shortcoming of Signal Us! was that the direct library space combined with only short instructions to customers limited the horizon of ideation. The expressed ideas veered toward incremental and immediate improvements. This was to be expected, and in the earlier lead-user workshop, this had been countenanced by intensively maintained future orientation to several years down the road. An unexpected issue in the Signal Us! day was the amount of support branch libraries needed in realizing it. What had seemed like very careful instructions and briefing by the participation planners were met with a considerable amount of clarification and instruction requests. When the event was on, it was found that enticing the customers to post their ideas required active interactions with customers unfamiliar to some of the staff members—engaging the citizens was not just about handing out Post-its. Further support was needed in storing and refining all the information received and software-based IdeaBank could have helped documentation and accumulation of insights in the face of the event that ended as a bit of a victim of its own success. And in all, this light participatory exercise was assessed to have triumphed over the traditional feedback gathering the libraries did and being also great in terms of organizational outreach and training to the wide range of library staff in active customer encounters. Within the library services Signal Us! was also deemed a success and repeated.

Friends of the Central Library: Setting Up a Citizen–Designer Community

The final collaborative design activity was the formation of a citizen design community to join the professionals in Oodi's planning to give participation a more continuous and institutionalized form in the Oodi project and beyond. Let us discuss this slightly more at length and in terms of the design participation work involved (see Chapter 2).

The community idea departed from three divergent ideas about a citizen–designer community: (1) the consultant (the author) introduced the community idea by emphasizing the win–win benefits of a permanent citizen–designer community, in keeping with user-innovation literature on hosted communities (e.g., Antorini et al., 2012; Prandelli et al. 2008); (2) the participation planner stressed empowering citizens toward equal and active engagement in design and dialogic relations in the planning of a major public undertaking, in keeping with participatory design (e.g., Dalsgaard 2012, 2013); and (3) the Oodi steering group foremost saw the community as a means of raising awareness for the Oodi project among citizens and elected city council members in keeping with representative democracy.

Even as these three points of departure were partially conflicting in the framing work they could be combined into five objectives: (1) testing existing ideas and preliminary service concepts with citizens; (2) resolving specific design questions with prospective users of those designs; (3) gaining new and unexpected service concept ideas; (4) bringing library staff and citizens closer together and introducing more customer-centered ways of working; and (5) using the results and processes in the development of the whole library network of Helsinki over the long haul. The objectives were further topicalized to peer learning, community-hosting, e-content sharing, and multiculturalism in Oodi's planning. The relevance work then continued into translating these topics into sub-areas that could be worked on by the participants and establishing the relevance of the initiative within the Oodi planning group as well as toward citizens, resulting in the Friends of Central Library (FCL) community concept, which was to be launched through online participation and a series of targeted workshops (Hyysalo et al. 2019).

The first collisions between democratic points of departure informing the FCL emerged in the selection work. FCL workshop participation was launched through a "job application" campaign in the media, which was viewed by 6700 people, resulting in 95 applications to join, with nice diversity among the applicants. But who should be chosen? The question was at once a principled one, regarding which ideals and ideas of democratic participation in design were followed in community composition, and also a constrained one, by simple practical issues. In terms of democratic principles, a public library is a tax-funded public institution for everyone equally, and thus, socio-economic-cultural representation was clearly an important principle. At the same time, the hackathon-type workshops might only work with citizens who had considerable domain expertise and design ability. A mix of representativeness and ability sounded good for community dynamics yet challenging to attain: the tension between representativeness and ability could not be resolved by the simple addition of more participants, as more people implied the need for more facilitators. Having four themes in each workshop meant that at least eight library staff members were needed as facilitators and notetakers to run the workshops. This number could not practically be increased much, and seven participants in each thematic group appeared as the manageable maximum, resulting in a total of 28 people selected for the workshop series with additional online possibilities. And finding good balance between representativeness and design ability among each 7 person group was considerably more challenging than a good mix among the 28.

Because of the cumulative progression of the three workshops, each step in the process more or less had to succeed in all subgroups, or else the next step could collapse (see Chapter 4). The use of four content themes meant, in effect, organizing 12 different mini-workshops for seven people instead of three events for 28 people. At the same time, the progression could not be very tightly scripted to foster participant ownership over goals, solutions, and the design process. The resulting eventual intermediate design work included staff training, workshop outlines, estimated progression, means, and facilitator roles and guidance on how to facilitate and record the discussions and solutions.

The collaborative design work began by acquainting the participants to the Oodi project and developing ideas in the theme areas. The ideas gathered from each group were numerous and of high quality, requiring only little polishing and clarification before being used to feed forward to the next workshop and to the Central Library main planners—FCL had started beyond all expectations! In the second workshop, the participants concretized their ideas into service concepts. Group dynamics and progression varied greatly between the four groups and influenced the group's capability to create ideas and refine concepts. Further issues arose from knowledge and skill asymmetries, as some participants were hacktivists or leading experts in specific fields (such as open data) or community activists (e.g., in a well-known youth squatters' association), whereas others were ordinary city residents. The solution to mix representative and design-savvy people proved tasking on facilitators, particularly in instructing the participants in the collaborative use of journey maps and blueprint templates. Nonetheless, the participants appeared very satisfied with the second workshop as well.

Yet, the templates and notes collected from the groups appeared scant, unclear, and incoherent, yielding little by the way of design concepts. What had happened? After a moment of disbelief and a touch of desperation, the participation planner proceeded to pursue "results archeology" and gleam from the facilitators and notetakers all the raw notes and service elements and asked them to explain them to her. This gradually revealed several original concepts and content discussions that had just been ill-documented in the flurry of the workshop. These concepts were presented back to participants in the next workshop, and (to organizers' great relief) the participants asserted that the concepts were very much what they had envisioned—just pending small edits and additions. The workshop then moved to find solutions to immediate open questions asked by architects and planners. Meanwhile, online participation was open to all citizens, and the online tasks received 5–40 comments/ideas each, some of which were further expanded upon in the workshops and integrated and published as part of the final results (Fig. 5.3).

In terms of output work, the yield of FCL start-up workshops was refined into design concepts in the seven areas, resulting in around 100 pages of documentation. It further validated and commented on numerous Central Library planners' ongoing design tasks and contributed critical insight that was divergent to planners' assumptions about the Central Library and its services. Both participant and in-house facilitator feedback was exceedingly positive. A designer–citizen community became a permanent fixture "library tribe" two years later, and FCL further contributed to the expansion of citizen participation activities in the City of Helsinki (Hyysalo et al. 2023).

In retrospect, FCL showed that institutionalizing participation in the citizen–designer community can be feasible and desirable among the citizens and public-sector organization alike. The setup phase showed that even resource-intensive in-person workshops could be viable, and at that, the community concept could cater for three divergent orientations to design democracy at once, straddling the academic and governance principles at the back of each of them (cf. Jensen & Petersen, 2016; Hyysalo et al., 2019). At the same time, the start-up phase workshops proved very skill- and labor-intensive for organizers. Various mundane practicalities such as the attainable participant and facilitator time and the labor required to refine the productions between workshops affected the workshop setup just as much as considerations of preferable forms of design democracy.

Fig. 5.3 Group ideation and discussion in one of the FCL subgroups (Photo: Virve Hyysalo, reprinted with permission)

Reflecting on the Oodi Participation Mix

In the Oodi project online and in-person idea gathering, workshops, and events for specific actor groups, collecting development ideas for existing libraries and pilots resulting from participatory budgeting and the citizen–design community (and so on) all utilized different formats of participation. These differed with respect to how many users could participate, how much say and design control the participants had, the timespan of participation, and with respect to the types of contributions and outcomes that ensued. For Library Services, these activities were by and large successful on their own and were deemed to have offered important complementarities regarding the people they were able to attract, the kind of design participation that took place, and the kind of information and design outputs that were gained. Each capitalized on different virtues of participation, such as offering a low investment way for undifferentiated publics to express their views and visions; allowing in-depth deliberations and decision-making; gaining insights on the thorny design areas; reaching out to ill-represented target groups of the future library; and complementing the host organization knowledge and participation gaps regarding what and how to design.

Some of the merits and downsides of design participation activities underscored in the post-mortem assessments related to the composition of the formats. It is not surprising that a considerably higher amount of filling in and registering information ensues from face-to-face interactions than from processing online idea entries. Conversely, it is easy to anticipate that online idea collection would have a higher propensity to result in terse and one-way statements in comparison to ones that were recorded through live dialogic interactions. Just as well, while the participants overall regarded the activities very positively, the ones who participated in the most in-depth and longest-running formats such as the citizen–designer community and lead-user workshop and maker pilot facility were the most satisfied. And quite predictably, while the relatively more in-depth formats gave most in terms of design substance, their contact price per citizen and stakeholder was manifold to the most fleeting and superficial participations. This naturally set boundaries as to how long the most intensive formats could be continued, such as the difference in how the FCL community was set up through a series of face-to-face events but was later converted to a largely online community.

At the same time, an important cross-cutting feature of all design participation activities and their post-mortem assessments was the themes elaborated on in Chapters 2 and 3: the activities were permeated by design participation work, and this affected greatly the yield of different participation formats. The greatest challenges remained with outcome work due to mundane limits to time to refine and analyse the outputs. This was also consequential, for instance, the weeks of work time that would have been required to carefully analyze all 2700 library dreams meant that only the most salient themes and categories were duly refined into thematic descriptions. These theme descriptions that were then available meant that they were the concepts that were taken into pilots project suggestions in the participatory budgeting. That they were piloted in turn solidified that some of them landed as features and services at the Oodi Library.

Regarding the permeation by strategizing and alignment work, Oodi design participation was intertwined with public relations and marketing, and a common gut reaction from many academics and activists would be that this likely diluted and compromised design participation. Yet in the Oodi preplanning phase, it turned out that the democratically most ambitious and new activities in the Finnish public sector, such as participatory budgeting and the citizen–designer community, were also best ones for gaining publicity. The prospects for gaining publicity for the project, in turn, allowed the use of a marketing budget and consequently a higher investment in participation in terms of participant recruiting and the numbers of people that could be involved. The prospect of publicity next led to emphasizing quality control, as any unprofessionally run events, failures in publicizing the results or acting on them could lead to public scrutiny and reputation loss. All this then affected which formats for participation were chosen, how the participation was conducted with these formats, and so on.

In summary, in the Oodi design participation mix, we see the important complementarities between different participation formats in reaching out to different people and in different ways and orienting to different outcomes. We also see the importance of the varied work of design participation—the work that the organizers, designers, and participants (must) do even though it goes beyond what was traditionally thought of as designing and beyond meeting any one set of ideals about public participation.

From Participation Mixes to Ecologies of User Knowledge

Let us now return to the starting point of the Oodi participation mix: the framing work exercise to identify who the publics are that are well covered with existing customer knowledge and which ones less so. While Oodi was a public innovation project to build a library of a new era, most its service areas and customer groups were such that Library Services already knew a lot about them to ground the planning. This is the usual case. Design hardly ever really begins de novo and neither does design participation. Previous design participation produces knowledge and engagement between the users and designers that prime future projects and other activities (e.g., marketing, organizational development), and, at best, this results in accumulation of insight on customers and publics in the organization that pursues participation, which is helpful for designing the next products and services.

At the same time, the historically formed ways of relating to customers may leave systematic blanks in knowledge about users and hinder the trials and appreciation of the yield of design participation approaches that could potentially be useful but clash with how things are done in the organization (Schrage 1999; Leonard 1995). There is thus a need to somehow take stock of what the organization already knows about its customers/users/audiences/publics and how it knows that; what participation formats may suit the organization; what are the groups of people and formats that might have been neglected, and just as well, what formats might not work because of organization's domain or its characteristics.

One might think such assessment would have merited plenty of attention in design management, open innovation or human-centered design, but, instead, it has remained a side-stream in several disciplines. In innovation management, it has been addressed through customer insight (Leonard-Barton 1995; Prahalad and Ramasvamy 2004; Piller and West 2014), but these models have remained generic, and the most contextualized models to date address user knowledge within an organization's absorptive capacity, which includes managerial goals and sub-capabilities, the underlying practices and method use, and interactions within those (Abrell et al. 2018), but as yet at a level of quite unspecified flow diagrams. In contrast, human-centered design, human–computer interaction (HCI), and information systems have created sophisticated models of customer organizations, but the assessment of the organization conducting the design has remained as yet as just clarifying its human-centered maturity (Jeffries et al. 2007; Earthy, 1989; cf. Hyysalo and Johnson 2015; Savolainen 2021). In these disciplines, it has been acknowledged that outside of academia, participation and user-insight methods are commonly mixed (Johnson 2013; Savolainen and Hyysalo 2020), but most efforts have still gone to validation of individual methods or combinations of two or three methods (e.g., Turnhout et al. 2014; Wardlaw 2016).

Science and Technology Studies has elaborated on how the matters involved are somewhat more complex in most real-life organizations. On the one hand, informal ways of knowing the customers and users have been shown to be part of almost all the design projects and organizations studied, affecting and complementing more explicit ways by which participation or research on users is conducted (Akrich 1995; Oudshoorn et al. 2003; Kotro 2005; Williams et al. 2005; Konrad, 2008; Pollock and Williams 2008; Hyysalo 2010; Wilkie 2010; Mozaffar 2016; Savolainen 2021). On the other hand, these studies make it clear that designers and design projects are hardly ever perfectly nested and aware of all the often-formidable knowledge and insight that the broader organization has accumulated (and partly forgotten) about customers and users (Hyysalo, Repo et al. 2016; Johnson 2013; Johnson et al. 2014; Jensen and Petersen 2016; Pollock and Williams 2008; Savolainen 2021).[4]

[4] Science and technology studies have spearheaded the detailed ethnographic studies, which show how real-life projects blend explicit studies on users with a range of informal and implicit understandings of users (e.g., Akrich 1995; Woolgar 1991; Sharrock and Anderson 1994) and that this work of representing users is an integral part of skilled design work (Kotro 2005; Heiskanen and Repo 2007; Jensen, 2012; Jensen and Petersen

Yet, despite these partial advances analytically depicting how different participation formats, existing customer insights, informal knowing (and false beliefs and stereotypes), and user research methods interrelate in an organization is challenging. Potentially the furthest it has been explored is the work on "ecologies of user knowledge" (Mäkinen et al. 2018; Savolainen 2021), which draws on a layered ecologies of knowledge mapping by Akera (2007). Akera's visual depiction allows examining how interrelated entities are arrayed to constitute a body of practice (Akera 2007, p. 415) through a layered representation without imputing pre-fixed causal relations and identities to sociotechnical entities.[5] The basic idea in this layered depiction of ecology of knowledge, or ecology of user knowledge, is to avoid the incredible complexity of interrelations that follows from flat ontology network depictions once they are deployed in earnest over the wide range of entities in large projects or large organizations, while not assuming a fixed status or hierarchy relations between the entities. The visual depiction, in turn, allows focus on the interrelations, which are clearer than a mere narrative description. For different uses and focalizations of the original ecology of knowledge mapping, see www.rpi.edu/~akeraa/articles/eK.

Akera's layers have been slightly amended to match the findings on user knowledge in organizations by Mäkinen et al. (2018). As the ecology of knowledge mapping does not impute particular a priori conceptions of sociotechnical order, further specification is merely an added means to make

2016; Savolainen and Hyysalo 2021). S&TS ethnographies further show that at least the mature organizations know a great deal about their users, and the issue is just as much about what potential insight and knowledge is ignored as it is about what insights are endorsed and integrated in a particular product or service development project (Hyysalo and Johnson 2016; Pollock and Williams 2008). Ignoring user insight is evident in terms of simply not being aware of what knowledge has accumulated within the organization, but it is equally evident in terms of judging the existing knowledge as irrelevant or misplaced because of its origins in different projects, products, eras, knowledge interests, or people in the organization (Johnson 2013). In sum, effective user insight in an organization consists of both informal ways of knowing the user and formally generated user studies, blended with existing stocks of previous studies, and is typically adapted piecemeal for new purposes and with new strategic considerations in mind.

[5] Ecologies of knowledge mapping should thus be taken as one of the means to "empiricize ontology" in practicee (Woolgar and Lezaun 2013; Marres 2012). It does not impose any pre-fixed sociomaterial topology on the empirics, yet it gives concrete means to examine what kind of ontological topology exists in the empirical study (Akera 2007). Ecological metaphors with complex causal determinants have long been used in symbolic interactionist sociology to emphasize the continuous mutual adjustment of actors (Star 1995) refraining from rigidly fixed system boundaries typically found in ecosystem metaphors in social sciences. Akera (2007) asserts that, in a body of practice, sociotechnical entities need to be arrayed in meaningful assemblages so that the coherence and meaningfulness of practice are retained. His mapping of an ecology of knowledge then uses the layers of actors, artifacts, knowledge/skills, organizations, occupations and disciplines, institutions, macroscopic institutions, and historical events as sensitizing concepts for tracing how a given body of practice is constituted.

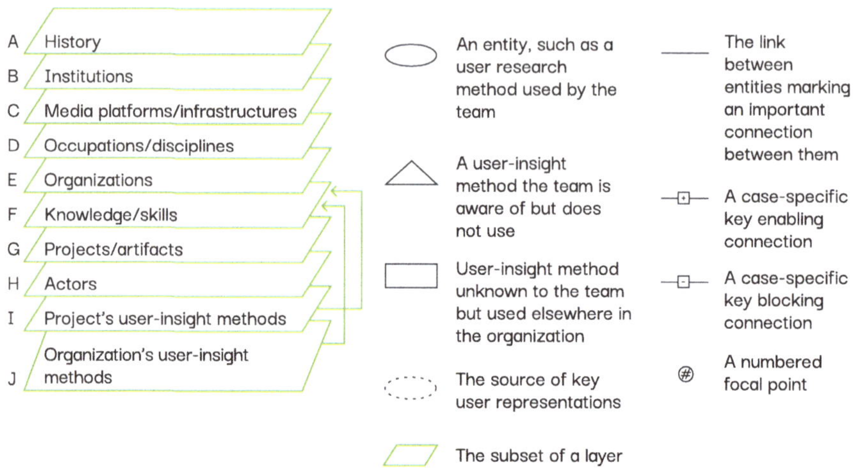

Fig. 5.4 The template and entity markers in mapping the ecology of user knowledge in an organization pursuing design participation (Mäkinen et al., 2018, reprinted with permission)

sense of how a specific body of practice has been sociomaterially organized. Akera uses several layers: history, institutions, occupations and disciplines, organizations, knowledge/skills, and actors, which are directly relevant for tracing user insight and participation efforts in an organization (Fig. 5.4). Mäkinen et al. (2018) alter the layer of "artifacts" to include projects because, in design, artifacts are commonly in-the-making. They further seek better focalization by expanding the skills layer into two, depicting the skills and methods in the design project and those available, in principle, in the whole organization. Also, to better focalize the mapping to participation and user insight, Mäkinen et al. (2018) split these layers minimally into three "settings": the participants'/users' setting, the developers' setting, and a shared setting in between the two, marked as layer subsets (cf. Muller and Druin 2009). They also add a layer of media platforms/infrastructures, as Mäkinen et al. (2018) were studying a media corporation where this was central. In their notation, key user representations were depicted as dashed lines, and they further introduced positive and negative connections between entities to better focalize the analysis of how methods and insights related in the design practices and further in the organizations. Their original notation is here further adjusted to work in black and white printing.

Ecologies of User Knowledge at the Finnish National Broadcasting Company

The mapping and its usefulness can be illustrated by the work of Mäkinen et al. (2018) analysis of two social media productions that took place at the Finnish

National Broadcasting Company (FNBC). FNBC is a large organization that, at the time, had over 3000 employees. It has a long tradition of producer- and journalist-led culture ever since its beginnings in the 1920s. The organization has sought a more customer-oriented approach in this millennium, resulting in a mix of older and new ways of relating to citizens ranging from, for example, audience research and focus groups to applied ethnography. Mäkinen et al. (2018) uncovered close to 70 customer research or user-involvement methods in total in their 55 interviews with FNBC staff, but only a few were in wide use.

A clearer picture of the ecologies of user knowledge emerges in concretely comparing two productions #lovemilla and learnweb2. The first one these was an online drama series, which was designed for youth aged 12–18 years old. The youth have historically been a difficult target group for the FNBC, but #lovemilla resulted in audience figures in traditional and social media that were unseen before at FNBC #lovemilla was a series of short episodes featuring a variety of different genres. It featured a mixed make-up from disguised educational topics to visits by real people well-known to its target audience. It was broadcasted on the FNBC's online TV platform and various social media, making it accessible 24/7. The main characters were "alive" on Facebook and Instagram—in which the target audience frequented at the time—and their profiles were updated by the project team.

How the #lovemilla production team related to its audience shows an unconventional mix that did not involve almost any formal design participation formats or customer research, merely some use of FNBC's corporation-wide audience data as a background resource. Instead, the team had extremely rich informal two-way interactions with the audience through social media in the design and deployment phases of the production (Fig. 5.5). From a human-centered design perspective, such customer insight would appear outright perilous as all rests on whatever informal insight was gathered, with very little analysis or validation.

Turning to the ecology of user knowledge mapping helps to clarify why these ways of knowing the audience were successful nonetheless (Fig. 5.6).

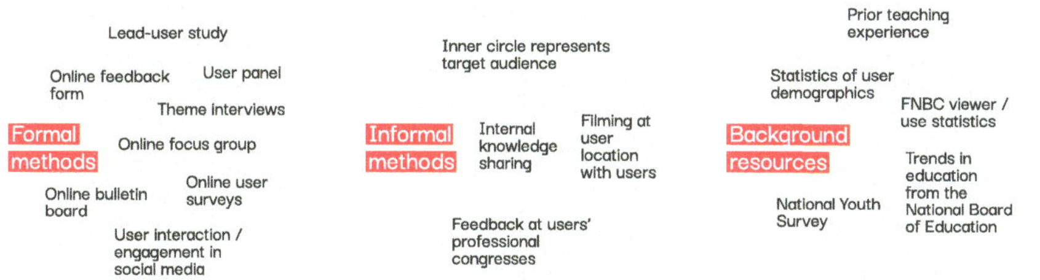

Fig. 5.5 Formal and informal methods and background resources used in the #lovemilla production

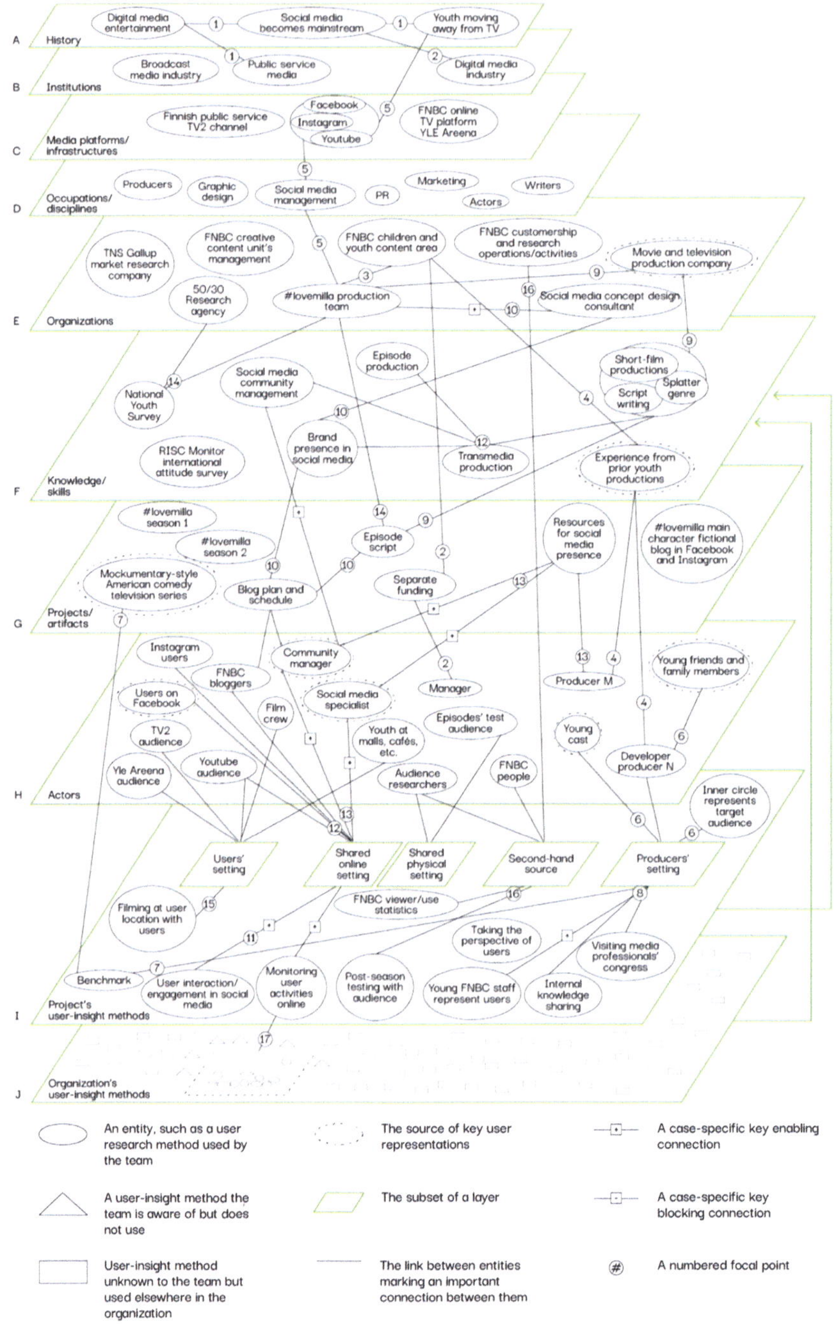

Fig. 5.6 User knowledge ecology of #lovemilla 2014. Numbers and letters referred to in the text appear in circles and layers. Adapted from Mäkinen et al. (2018)

While FNBC had a long history of successful children's productions (B1), it had a less consistent record with youth who were also increasingly moving to digital media and away from TV (A1). In response, the management of FNBC (H2) granted a separate small in-house funding to initiate #lovemilla (G2), resulting in a small but flexible and independent production team (E3). The producers (H4) had prior experience with episode scripts and building dialogue (F4), and this underlined the importance of staying in close touch with the target group on major social media platforms—at the time Facebook, Instagram, and YouTube (C5)—to the extent that a social media producer (D5) was deemed a necessary and vital part of the #lovemilla production team (E5). This formed the basic setup in which the #lovemilla concept took shape. The producer benchmarked the field of episode broadcasting (E7), including American mocumentary comedy (G7), visited trade events in the field, and took part in the FNBC internal exchange sessions. The producer also sought the youth perspective actively, and their little sister represented the youth as a target group (H6, I6), while the production worked actively with FNBC's young staff members. All this happened at the "producer's setting"—the domain of professional media production (I8). These resources led to the basic concept.

The team next hired a television production company (E9) specialized in short films and splatter movies (F9) and gave them the task of developing the scripts for the episodes (G9). They further hired a social media consultant (E10) to take care of #lovemilla's social media presence (F10), blog plan, and schedule in conjunction with the episodes (G10). The social media interactions (I11) created a shared online setting between users and developers (I12), in which the in-house community manager operated (I13), made possible by specifically allotted funds (G13). This was vital as it made two-way interactions and co-acting possible in addition to the blog schedule, enabling collaborative content and community management (F12). The episode scripts (G14) were mostly developed via informal insights in the producer's environment (I8, I6) and filming youth programs at users' settings (I15), with some recourse to background resources such as the National Youth Survey (F14).

The visual mapping shows how the #lovemilla production utilized informal knowing by its team (I8, I15) paired with two-way interactions in social media (I12) as inputs to the transmedia production and as a part of the format itself. This was supported by interactions with some surrogate customers standing in for the audience. In contrast, the organization's user-insight methods were not deployed (J17, E16) or were reduced to secondary complements (I16). This concentration on two-way interactions and informal knowing supported the fast-paced production that had a remit to trial and implement ideas quite freely. We thus see an alignment with the FNBC producer-oriented history updated onto fast-changing youth interactive media, with very few blocking or hindering connections.

#lovemilla's ecology of knowledge can, in a sense, only become appreciated when contrasted to another project in the same organization that sought a much more "by the book" approach to human-centered design and user involvement. Learnweb was an online service launched in 2007 with the main purpose of offering useful educational materials for teachers that could be shown in class. It gradually shifted orientation toward being a repository of learning-related materials to the broader public. FNBC's strategy emphasized participation by the audience, and Learnweb2 utilized a whole suite of formal user-involvement and research methods, complemented by background information and informal methods, while limiting its social media engagement to the possibility of posting content and sharing it on major social media platforms (Fig. 5.7). Despite these formidable efforts, the audience numbers remained lower than expected—thus, in effect it formed a polar opposite of how #lovemilla developed.

The ecology of user knowledge mapping (Fig. 5.8) again helps to fathom why. FNBC had plenty of educational content in its TV channels, which the initial Learnweb (G1) opened on a digital platform (C1) in response to a gradual historical change of roles that digital media and TV played in schools (A1). The service was half-funded by the National Board of Education (G2), and this liaison between two public organizations needed to steer clear of commercial educational publishers' digital platforms (C3) to avoid distorting competition by publicly funded services (B3).

The Learnweb development team (E4) was organizationally located in a unit that was an early mover in social media (F4) accustomed to still relatively one-way communication of bloggers (H4) posting content (C4) and, for instance, not having a separate social media manager to spur interactions. Audience participation took place in other ways, such as teachers coproducing content when programs were filmed at schools (I5) and receiving feedback through an online form and online bulletin board, a focus group, and online surveys for the panel of about 100 teachers (I6) and the help of audience researchers (H6). Theme interviews (I7) provided additional insights as did the team members who had teaching experience (F8) and were in tune with professional trends and teachers (I9). The National Board of Education provided periodic updates on changes in education (F10). The project thus

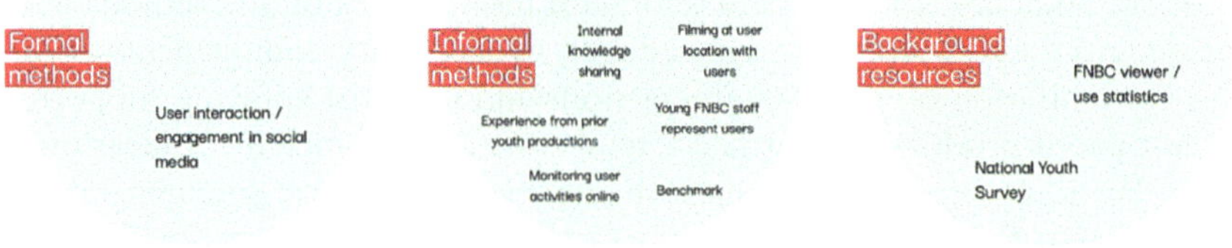

Fig. 5.7 The formal methods, informal methods, and background resources for the Learnweb2 project

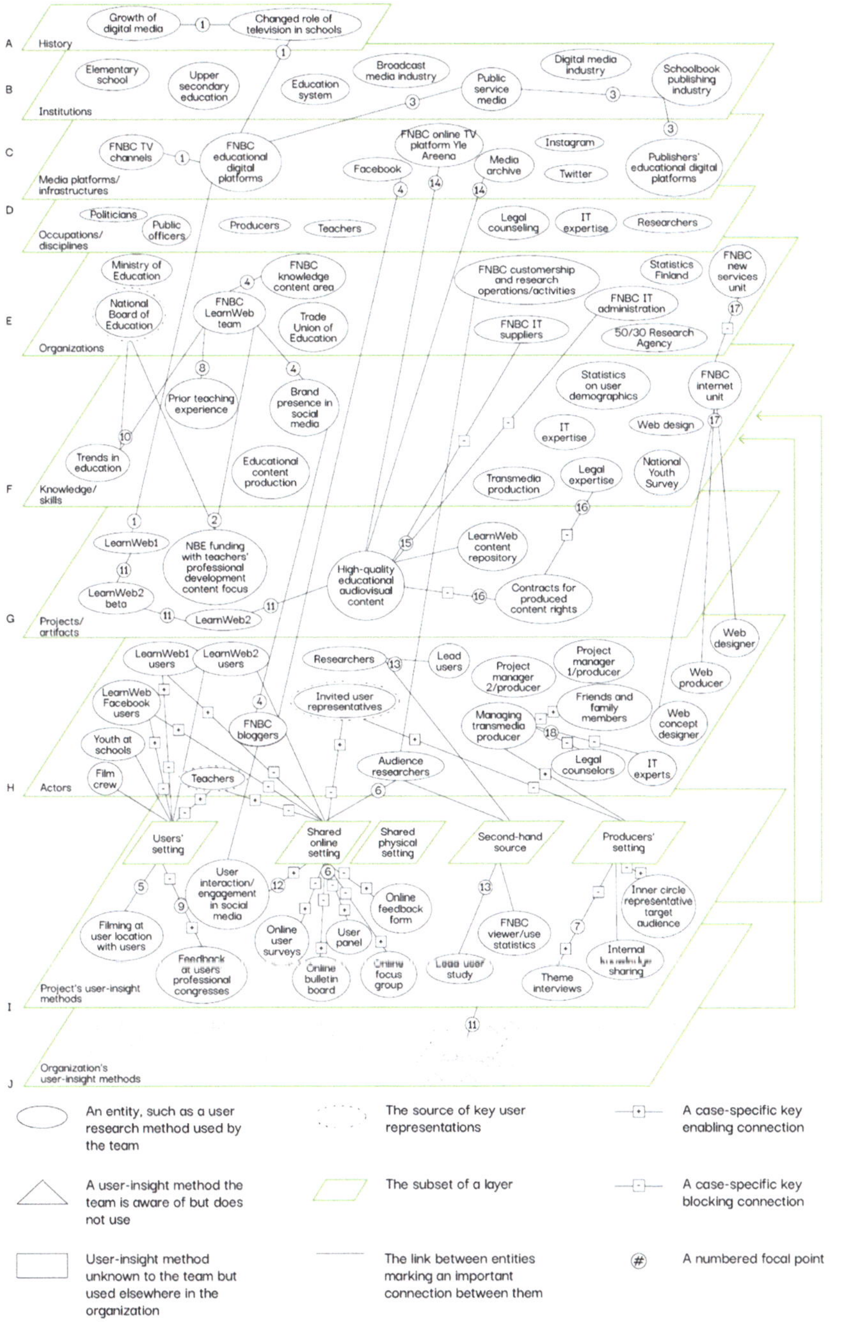

Fig. 5.8 User knowledge ecology of Learnweb in 2013. Numbers and letters referred to in the text appear in circles and layers

utilized a wide array of the organizational repertoire for customer engagement and audience insight (J11) with limited informal social media interactions (I12).

Yet the Learnweb2 project faced ambiguities about its target groups and their changing needs. It was thus selected for a pilot lead-user study (I13, H13) on online education solutions (Hyysalo et al. 2015; Hannukainen et al., 2018). It also expanded its audience from teachers to the general learner populus in keeping with FNBC's strategy and sought to integrate all its learning materials and other documentary repositories (C14), which proved arduous even with an external IT supplier (G15) and was further hindered by varying copyrights related to old materials that did not cover things such as online sharing (G16). Further rigidity arose from the liaison between FNBC and the National Board of Education (G2) and FNBC's restructuring (E17, F17), in effect curbing resources for online development (H18) and preventing the resolution of intermittency problems in the projects' staffing.

In all, the mapping displays an impressive deployment of design participation formats and ways to engage with the potential users. At the same it displays the organizational and historical ties affecting the Learnweb, which were needed to make accountable decisions, pushing it toward more formal methods and user research (J11), while holding limited room to experiment and act upon customer insights. The predominantly one-way and limited use of social media and the production itself as a setting to interact with participants becomes understandable through the team and organizational and inter-organizational histories. The number of constraining and constraining-enabling relations in the Learnweb's ecology of knowledge makes it evident that the substantial amount of design participation and user insight formats likely was needed to substantiate action, but even this thoroughness could not quite compensate for the need to act and experiment in the rapidly changing field of online educational resource use and provision.

Chapter Conclusions: Reflecting on Participation Mixes and Ecologies of User Knowledge

Participation mixes and ecologies of user knowledge highlight the need to contextualize design participation adequately among the participating people and within an organization that seeks to engage people in it. Rather than seeking to shoehorn participation into any one endorsed method or wider format of participation, an assessment ought to take place of the matters involved, the existing and emergent actor groups' publics, and the already accumulated insights and pattering of previous design participations. This commonly results in the realization that a participation mix ought to be pursued to adequately cater to the different aims and virtues of participation.

Such assessment can be pursued informally or with light visual structuring as we did in the Oodi case. Yet the more formidable analytical sensitivity and template provided by user knowledge ecology mapping may help in the process. This is particularly the case whenever the design participation is taking place in a context where a lot of insight about the participants already exists. Ecology of user knowledge mapping can guide the assessment and display of relevant entities and connections within the team and toward its patrons and stakeholders. The layered structure of the template also helps multidisciplinary teams in which some members are not familiar with organizational knowledge flows or human and social science processes, as well as in communicating the often complex and uncertain status of insights from design participation and user research.

The cases all underscore the need to assess complementarities between design participation formats in conjunction with hindering facets and the capacities to act on the likely outputs and insights. Minimally, the ecology of user knowledge mapping should act to counter the untenable assumptions that the more in-depth the participation is and the more explicit the user research is the better. The ecology depictions also help to indicate why some sources of insight about future users become effective and others do not, giving more concreteness to what works and why in "methods mixes" and "human-centered capacities" (Johnson et al. 2014a; Abrell et al. 2017), helping to address the crucial questions of applicability in different design participation formats within the actual organizational constraints.

In all, for the wide uptake of design participation, considerations of participation mixes and ecologies of user knowledge facilitate reflection and learning about enabling and blocking connections in the ways participation is organized. A more critical and strategic analysis can also focus on why some methods end up being used widely across projects: is it a matter of happenstance, familiarity, and legacy or do these methods have characteristics that are particularly advantageous or compatible with the organization's processes or its customers and stakeholders? In all such considerations, the idea of participation mixes and ecology of user knowledge mapping help render decisions about design participation into a more informed basis.

CHAPTER 6

Zooming Out to Map Participation Approaches Through Producer–User Configurations

Introduction: Making Sense of the Multitude of Methods and Approaches to Design Participation

This chapter "zooms out" to examine the key differences between the approaches to configuring designer–user relations for design participation.[1] We shall first recount the currently available descriptions, listings, and mappings of user, customer, and participant involvement and elaborate on the persistent conceptual shortcomings that have hampered attempts to order different methods of design participation. We then indicate one way by which to differentiate between families of approaches to design participation with hopefully better conceptual clarity. While this may sound like a very academic pursuit, it has tangible practical implications: each approach family holds different import for the roles of designers and participants and the primary ways by which participation is configured and, at this, holds different strongholds and weaknesses. This makes it practically important for those pursuing design participation to understand what is involved in one approach or another, as well as in mixing and shifting between them.

As noted an increasingly wide array of arrangements—methods, tools, procedures, techniques, and methodologies—has emerged by which users' practices, knowledge, creativity, and efforts can be engaged with (e.g., Sanders and Stappers 2008; Hyysalo, Jensen et al. 2016; Galabo et al. 2021). New application areas, digital services, and platforms have enabled new ways to

[1] This chapter has been extended and edited for the present book from an article originally published under CC-BY 4.0 license: Hyysalo, Sampsa & Johnson, Mikael. (2024) Making sense of methods and approaches to user involvement. *The Design Journal*, 1–29.

S. Hyysalo, *Design Participation*,
https://doi.org/10.1007/978-3-031-95437-5_6

organize design participation across time and space, sharing knowledge and pooling contributions among users and between users and designers (Bovaird 2007; Botero and Saad-Sulonen 2010; Voss et al. 2009; Leminen 2015; Hyysalo, Jensen et al. 2016; von Hippel 2005). This multiplication of forms for design participation has created difficulties for practitioners, students, and academics in keeping pace with what differences exist in the different ways users are involved or invested in and for design.

Our research on real-life projects in private and public organizations and citizen communities indicates that practitioners use methods they happen to be acquainted with or encounter. In so doing, practitioners tend to neglect possibilities that require ways of working that are unfamiliar to them (Hyysalo, Repo et al. 2016). To give a concrete example, it is common for design practitioners to equate codesign with running collocated workshops, and if that remains the case, it is then those that become carried out even if more distributed or asynchronous forms of participation might have suited the project context better (cf. Chapters 3, 4, and 5; Hartswood et al. 2002; Karasti et al. 2010; Botero and Hyysalo 2013; Botero et al. 2020). This underscores the need to better elaborate differences in design participation across the continuum of their involvement from mere informants to independent designers (Arnstein 1969; von Hippel 2016; Hyysalo, Repo et al. 2016).

As noted in Chapters 1 and 2, methods have occupied an exceedingly central place across human-centered design (HCD), participatory design (PD) and community-based design and innovation. Across these areas, there are both (sub)field method listings and "mixed origin mappings" or "landscape mappings." Even as we are rather critical to the sufficiency of this methods orientation on the whole (see Chapters 2–5), the methods continue to hold such a central position in the academic and practitioner discourse that one should hope there were at least productive and well-differentiated ways to discuss methods, their differences, and likely yields and shortcomings. This is where we turn to next by recounting how methods have been differentiated in the literature to date and then moving on to provide a more robust alternative based on this analysis.

The Mappings of Methods in Design Participation

Let us depart into the methods of design participation from subfield-specific method and technique mappings. A good example is the influential mapping of PD methods and techniques that Muller and Kuhn (1993) wrote already in the early 1990s and that has been reprinted in several handbooks since. They organize PD activities according to two dimensions. The y-axis differentiates between creating knowledge of the use context and directly drawing users into design. The x-axis position signals whether the methods and techniques are applicable early or late in the (imagined) development cycle. They further indicate the geography, use in industry, and suitable group size. of design participation from subfield-specific method and technique mappings. A good example is the influential mapping of PD methods and techniques that Muller and Kuhn (1993) wrote already in the early 1990s and that has been reprinted in several handbooks since (Fig 6.1). They organize PD activities according to two dimensions. The y-axis differentiates between creating knowledge of the use context and directly drawing users into design. The x-axis position signals whether the methods and techniques are applicable early or late in the (imagined) development cycle. They further indicate the geography, use in industry, and suitable group size. of design participation from subfield-specific method and technique mappings. A good example is the influential mapping of PD methods and techniques that Muller and Kuhn (1993) wrote already in the early 1990s and that has been reprinted in several handbooks since. They organize PD activities according to two dimensions. The y-axis differentiates between creating knowledge of the use context and directly drawing users into design. The x-axis position signals whether the methods and techniques are applicable early or late in the (imagined) development cycle. They further indicate the geography, use in industry, and suitable group size. of design participation from subfield-specific method and technique mappings. A good example is the influential mapping of PD methods and techniques that Muller and Kuhn (1993) wrote already in the early 1990s and that has been reprinted in several handbooks since. They organize PD activities according to two dimensions. The y-axis differentiates between creating knowledge of the use context and directly drawing users into design. The x-axis position signals whether the methods and techniques are applicable early or late in the (imagined) development cycle. They further indicate the geography, use in industry, and suitable group size.

While this mapping is instructional, the fixed connection of a method or technique to a specific position—early or late—in a development cycle or iteration has by now become debatable because it presumes two assumptions that are true only for some development projects. In this mapping, and many other similar mappings, the first common assumption is that developers and designers know little about the users early in a project. The second is that the product is so immature that it cannot be user tested at an early stage. This image of the distanced designer working with ideas and raw materials fits blue sky innovation and the designers being academics or consultants being deployed into a context that is new to them. In contrast, many real-life projects continue some earlier development efforts, which implies that designers can lean on their experience from prior projects and that there already is a service or product in production (Johnson 2013; Johnson et al. 2014a, b; Savolainen 2021; Mäkinen et al. 2018).

In contrast to potentially more "user knowledge poor" organizations of the early 1990s when Muller and Kuhn (1993) made their mapping, in many of today's development contexts, knowledge about users is cumulative and combined, and many projects can benefit from real user data already in the beginning of the project (see Chapter 5; Savolainen and Hyysalo 2020; Savolainen 2021). This is particularly underscored in "perpetual beta" digital

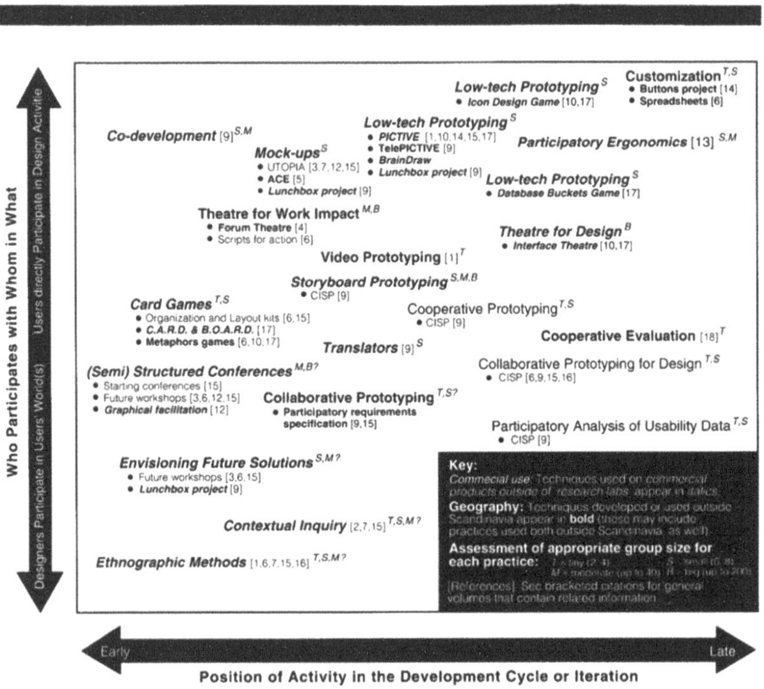

Fig. 6.1 A mapping of participatory design methods (Muller and Kuhn 1993). Reprinted with Permission

service development and other design-in-use time approaches, in which the development of the same system is continued through many rounds of iteration (see Chapter 3). Because of the broad variance in real-life projects, the development position or project phase is difficult to stand on as a generic way of grouping methods, except for instructional purposes and a limited range of projects—as one can observe in Table 6.1 and Fig. 6.3, its use is nonetheless very common in various methods categorizations.

The subfield mappings need scrutiny also with respect to the grouped items, so let us discuss a few other subfield mappings to clarify this. Subfield mappings are not specific to PD, and for instance in HCD, we can find methods mapping such as the usability net (Maguire 2001; Table 6.1). This kind of mapping allows authors to compare the relevance, merits, and limitations of methods (Maguire 2001; Kaulio 1998; Kujala 2003). These within-field mappings have recently developed into electronic resources. For example, in user experience (UX) design research, www.allaboutux.org provides resources and listings of 82 UX techniques, methods, and methodologies with search functions that in some repositories allow entering basic project parameters as well. Similar to www.ucdtoolbox.com (van Kuijk and Staats 2012), actioncat alogue.eu maps stakeholder engagement methods and techniques with basic project parameters.

Table 6.1 A listing of human-centered design methods (Maguire 2001)

Planning (section 3)	*Context of use (section 4)*	*Requirements (section 5)*	*Design (section 6)*	*Evaluation (section 7)*
3.1 Usability planning and scoping	4.1 Identify stakeholders	5.1 Stakeholder analysis	6.1 Brainstorming	7.1 Participatory evaluation
3.2 Usability cost benefit analysis	4.2 Context of use analysis	5.2 User cost benefit analysis	6.2 Parallel design	7.2 Assisted evaluation
	4.3 Survey of existing users	5.3 User requirements interview	6.3 Design guidelines and standards	7.3 Heuristic or expert evaluation
	4.4 Field study/ user observation	5.4 Focus groups	6.4 Storyboarding	7.4 Controlled user testing
	4.5 Diary keeping	5.5 Scenarios of use	6.5 Affinity diagram	7.5 Satisfaction questionnaires
	4.6 Task analysis	5.6 Personas	6.6 Card sorting	7.6 Assessing cognitive workload
		5.7 Existing system/ competitor analysis	6.7 Paper prototyping	7.7 Critical incidents
		5.8 Task/ function mapping	6.8 Software prototyping	7.8 Post-experience interviews
		5.9 Allocation of function	6.9 Wizard-of-Oz prototyping	
		5.10 User, usability, and organizational requirements	6.10 Organizational prototyping	

While these "within-subfield" listings are useful, they tend to close the space to neighboring disciplines in a way that is hard to justify in terms of genealogy and unity of methods. For instance, in Table 6.1 we find a wild variety of old techniques and origins: *focus groups* were invented by sociologist Robert Merton in the 1950s for quite a different purpose than design; *stakeholder analysis* emerged between the 1930s and 1960s for use in corporate responsibility; *brainstorming* was derived from marketing creativity exercises after World War II; and *card sorting* in work modeling evolved within PD and development studies in the 1970s. How these methods are the most relevant to human-centered design remains a small mystery, similar to how many methods belong in Muller and Kuhns's (1993) PD method mapping, i.e., why these methods and not others "belong" to such a seemingly "field"-specific mapping. And, as importantly, does the use of these methods imply HCD or PD even though many of the methods are used widely and even predominantly in other fields?

As the genealogies reveal, the "fields" that delineate methods to be mapped can be rather arbitrary constructions. Most listings are factually "mixed origin mappings," which some of the authors admit from the outset (e.g., Muller and Druin 2012).[2] This is the case, for instance, with the Danish user-driven innovation portal listing www.innotour.com/innovation-tools/user-driven-methods/ and "landscape"-type mappings, such as Steen (2011) and Sanders and Stappers (2008). The key merit of such mixed origin/landscape mapping is that it recognizes emerging areas in the landscape and seeks to organize the mutual relations between different entities. At best, its organizing principles recognize that different ways to organize user involvement may entail different design processes and thus do not lapse into a problematic unified design process as a backbone (as critiqued above).

Let us examine an attempt at tracing the genealogies of methods and acknowledging that they have been developed in multiple disciplines. We are sympathetic to this idea as well but regret that it may not work too well in practice. www.servicedesigntools.org provides, among its other resources, a map of service design tools, which is too big to be represented in this book, but it can be found online here: https://servicedesigntools.org/assets/images/uploads/tools_provenance.pdf. The map aims to demonstrate how service design tools evolved over time and their provenance, and it nicely indicates that some methods commonly discussed in design can be originated in social sciences and marketing, as well as in engineering design and operations management. It proceeds to show these influences via hierarchical ancestry tree presentation—with roots, branches, and leaves—clearly borrowed from cladism in biology.

As appealing as cladism may be, when applied to design it leads to a considerable mapping muddle. First, branch maps overemphasize divergence over time as convergence and crossover adoption is not incorporated (even as it featured widely in the above examples, see Table 6.1 and Figs. 6.1, 6.2, 6.3). This makes mixed origin methods, tools, and techniques impossible to represent appropriately, since it would require that the same leaf belongs to several trees. For instance, let us consider the tool that www.servicedesigntools.org calls *tomorrow's narratives*, which is a generic technique to envision the future in order to elicit a vision. Very similar future-oriented design techniques exist in parallel fields, as well as in *user experience, usability, and participatory design*. Are they now all wholly different just because they borrow from future studies to different design fields? Second and related, the map makes dynamic changes

[2] For a more sophisticated way out of this quandary, see Muller and Druin (2012), who elaborate on how many methods can be rendered participative by using them collaboratively in the interaction between designers and users, often requiring some modification to suit this "third space" usage.

and exchanges within and across fields difficult to represent. For instance, *user experience* is placed as starting already in the 1930s, which we conjecture potentially owing to the first referrals to customer experience that occur at that time in marketing, when, in contrast, user experience is commonly referred to as a phenomenon that emerged in the mid-1990s within *human–computer interaction* (HCI), in which it expanded and extended the field of *usability*. Another example, *experience prototypes*, which is a tool that is used to simulate the service experience by prototyping, is simultaneously presented as emerging from the *user experience* root in 2000 (also called *low-fi prototypes* and *mock-ups*). This could indicate that the idea has "traveled" from *mock-ups* from PD in the 1970s, via usability to user experience in the 1980s–1990s, and then embraced by service design. But such complex routes of travel and cross-fertilization are not represented, and they would require some other typological principle added on top of the tree branching to become successfully mapped.

Sanders and Stappers's (2008) famous landscape mapping, as a "cognitive collage" of HCD research, avoids these genealogical pitfalls (Fig. 6.2). They structure the items into what they call zones, clusters, and bubbles. For example, in the large user-centered design zone, we find three clusters (human factors and ergonomics, applied ethnography, and usability testing) and two bubbles (contextual inquiry and lead-user innovation). They position the items on two axes ranging from the user as a subject to the user as a partner and led

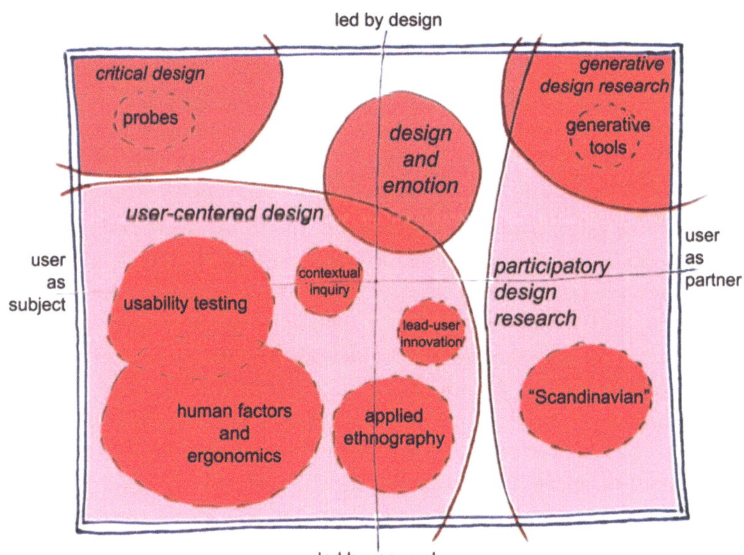

Fig. 6.2 A landscape mapping of codesign (Sanders and Stappers 2006, 2008). The later iterations of the same mapping proceed with the same basic schemata in a simplified manner. Reprinted with Permission

by research and led by design. Such landscape mapping has the potential to better address the origins and differences between different methods as well as to avoid problematic constructs such as applicability to assumed phases of design.

Yet, seeking to organize wide and arguably quite a motley collection of approaches becomes tricky. Granted that Sanders and Stappers (2008) make a caveat that their mapping is just *their cognitive collage*, we can push this a little and consider what it might take to make the practitioners and researchers implicated by the mapped items to duly recognize themselves in the mapping, as this then raises to the fore a set of key issues that all the other available mixed origin and landscape mappings struggle with as well.

The first problem area would concern *item structure and naming*. On what basis are the items cast into their places, such as zones, clusters, and bubbles in Fig. 6.2? The four zones (critical design, design and emotion, user-centered design, and participatory design) could refer to approaches to design research, since the authors state that "participatory design is an approach," but the "approach" term is used ambiguously. For instance, both *lead-user innovation* and *generative tools* are also denoted as approaches, even though they are "mere" bubbles. This segues nicely into the second common problem area, which is precision needed in item naming. For instance, *lead-user innovation* is not a category by which the close to 1500 researchers focusing on *innovation by users* identify themselves. Rather, *user-innovation research* comprises research on a considerably broader array of topics and users than just *lead users*, even if one of its main findings has been that innovation by users tends to be concentrated on people that meet lead-user characteristics (von Hippel 1976, 2005, 2016). If innovation by lead users was only a naturally occurring phenomenon and not closely connected to a particular stream of research, such naming could be okay if no other good "native" term existed yet. Importantly, the relationship between existing terms and mapping terms raises an even more important issue, which is *item extension and status*. For instance, the Fig. 6.2 mapping gives *lead-user innovation* and *contextual inquiry* an equal status, which confuses their relative position within the methodologies and approaches they are part of. Within user-innovation research, there are several methodologies, and if one chooses, for instance, the *lead-user method*, then the counterpart to *contextual inquiry* (Holtzblat and Jones 1993) would be the *lead-user identification process*. In turn, the *lead-user method* (von Hippel 2005; Churchill et al. 2009) would be comparable to *contextual design* (Beyer and Holzblatt 1998). Nested relations pose further delimitation issues, such as in the cluster of applied ethnography within the HCD zone that uproots ethnography from its academic and professional history of over a century in both anthropology and social sciences at workplaces and, by now, many kinds of applications in marketing, innovation, and organizational reform. Because of this, its "applied" nature in the "User Centred Design zone" could point to several different ways it has been applied independently as well as part of design methodologies.

In all, by discussing in detail recurring problems in "mapping" methods and approaches, we hope to have made it clear that such mapping is not at all easy to do, even if one pretends there is a subfield island or whether one acknowledges that there is an almost inbuild need to address the broader landscape.

Now designers sometimes voice that the criterion for how user-involvement methods interlink should be the way by which professional designers see their differences and how they view them in their practice (Law et al. 2009). This avenue of thought may work well in clarifying how practitioners perceive or use methods, but it may sound better than what it can actually deliver once it is applied into methods taxonomies. An example of how professional designers' perceptions have been formalized to arrive at a methods map is the card-sorting exercise with designers by Goodman-Deane et al. (2008). They identified a large number of design methods and techniques covering fields such as product design, HCI, and ergonomics. Then they chose 57 methods for understanding and involving users aiming for a representative range of method types. Each method was described on a card, and 21 designers organized the methods into groups as they wished and labeled the groups. Based on a cluster analysis, they ended up with the method items and groups in Fig. 6.3.

While the work by Goodman-Deane et al. (2008 is instructive in showing how mainstream designers thought at the time, it under-represents the available methods for user participation and favors indirect user representations (Hyysalo and Johnson 2015). Direct user contact (cluster F in Fig. 6.3) is limited to interview-based techniques (*talking to users*, such as *codesign* and *focus groups*) and non-interview techniques (*other user contact*, such as *shadowing* and *usability testing*). This puts "codesign" in an odd place, since in contrast, from a codesign perspective, one could argue that users could be present and participate in any of the A–F method clusters. Indeed, professional designers' categorizations may in fact be rather biased, conservative, and potentially more—not less—confused regarding user involvement than the academic attempts at clarification.

Thus, to sum up this section, clarifying the differences and relations between methods and what can and what cannot be sensibly mapped calls for more careful typological work.

Clarifying the Layers of Approaches, Methodologies, Methods, and Techniques

Some parameters needed for better mapping methods and approaches to design participation can be found in the neighboring fields, namely methodology literature in information systems and HCD. Information systems development (ISD) methods have been decomposed into four layers by Iivari et al. (2000). Their most incremental layer is *techniques* where they list items, such as use cases and rich pictures. Their next layer is *methodologies,* which

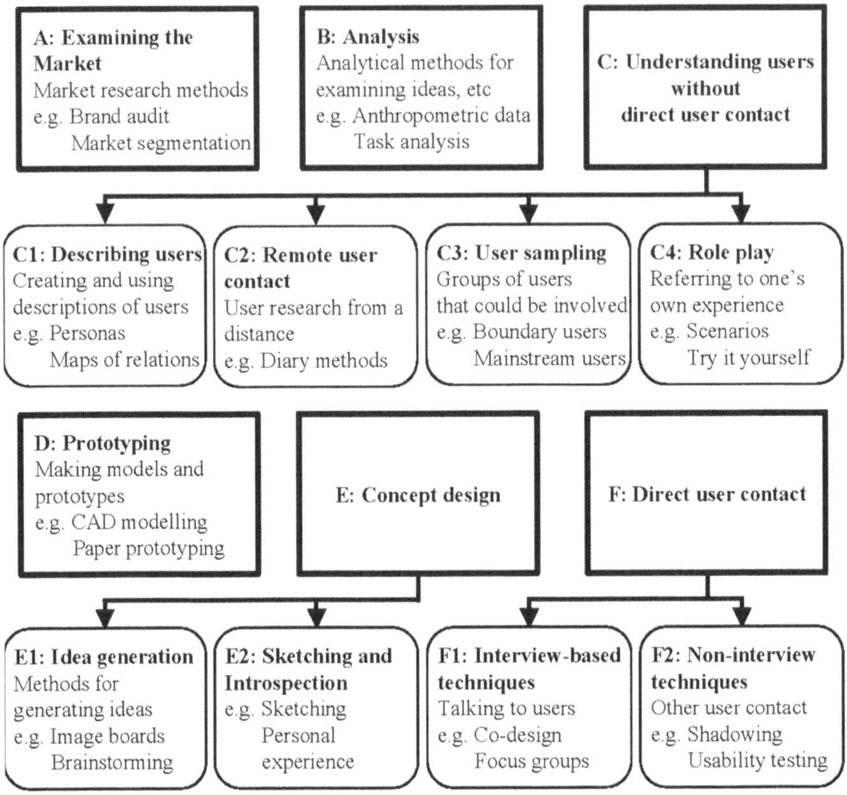

Fig. 6.3 Clusters of methods (Goodman-Deane et al. 2008)

consist of a combination of techniques and establish the relations between them as a detailed development process. The methodologies, in turn, are seen as *instances* of ISD *approaches* that hold more overarching goals, guiding principles, fundamental concepts, and principles for the development process. Their final layer is ISD paradigms, which they take approaches to instantiate[3] (Fig. 6.4).

Similarly, in usability, Woolrych et al. (2011) clarify how "methods as recipes"—such as cognitive walkthrough or user testing—are factually complemented by what they call the "resources" that go into realizing the method

[3] A considerable proportion of methods for designing for, with, and by users is not articulated to the level of scientific (proto-)paradigms, which Iivari et al. (2000) model after Burrell and Morgan's (1979) paradigms in organizational theory. Moreover, sociological paradigms for organizational analysis do not adequately characterize the underlying assumptions in collaborative design approaches that tend to be more mixed and practice-originated. There are also considerable problems in Burrell and Morgan's delineation of their paradigm constructs which they themselves have denounced later. But the rest of Iivari et al. (2000) reasoning is very useful.

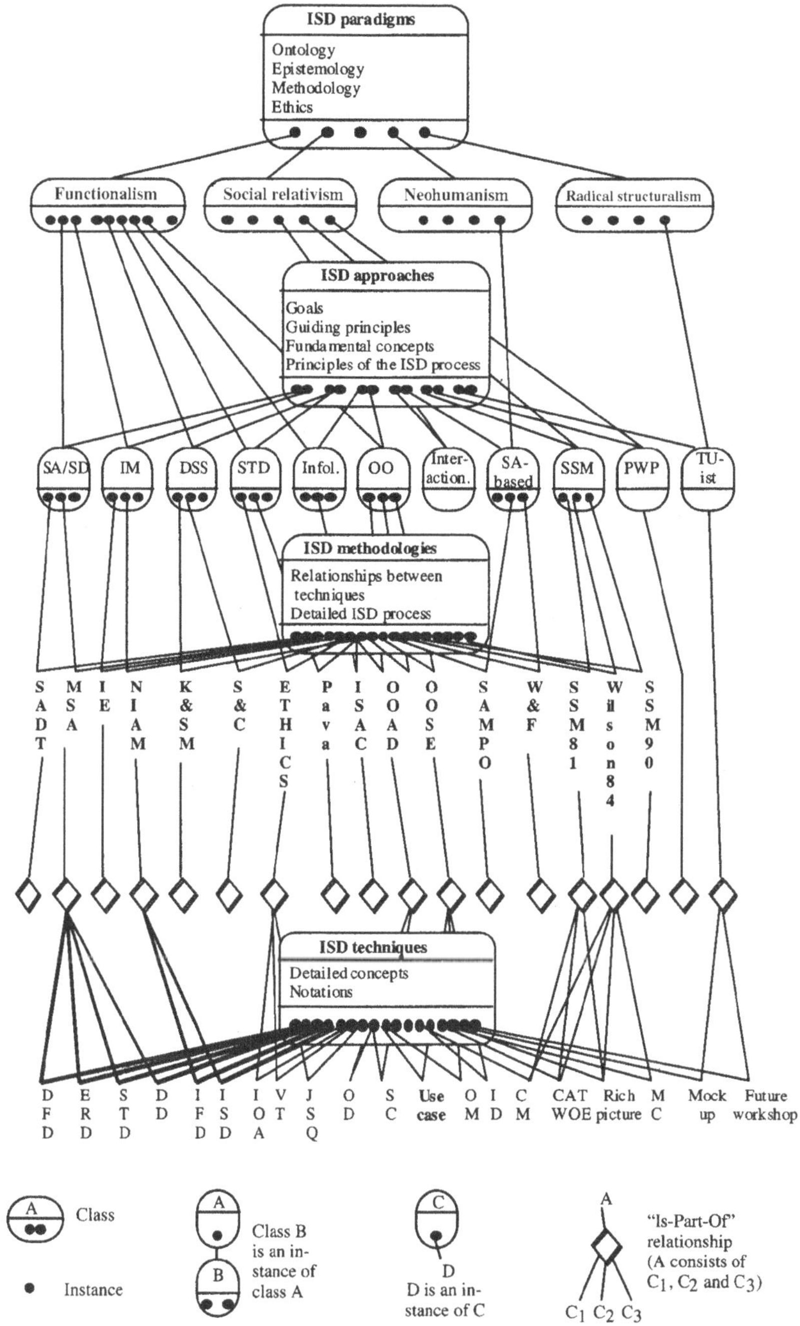

Fig. 6.4 The hierarchy of information systems development paradigms, approaches, methodologies, and techniques (Iivari et al., 2000). Reprinted with Permission

recipe, such as enrolling the participants, reporting formats, merging findings, and the detailed carrying out of protocols. They further point out that "method recipes" are hardly ever used but are used as part of "usability work" in a particular development context and have been factually found to include "method-mixes" (Johnson et al. 2014a). Three layers are thus discerned, those of "usability work," "methods as recipes," and "practical tasks and resources."

What these works together suggest is a typology of four layers that we find helpful in clarifying the interrelations of different items for design for, with, and by users. The layer of methods as recipes by Woolrych et al. (2011) and the techniques by Iivari et al. (2000) have, by and large, the same extension, which is clearest to call *methods* (such as Future workshop). Woolrych et al. (2011) decompose these methods further into practical tasks, representations, and resources that denote the more elemental end of the techniques by Iivari et al. (2000) and which is clearest to be called *techniques*. These are needed to realize methods—that is, methods consist of the application of techniques and, in turn, *methodologies* consist of a combination of methods to achieve broader ends.

Finally, methodologies are *instances* of families of approaches to how designer–user relations have been organized in design and what the predominant user-involvement configuration is in them, such as collocated participation in traditional PD, investigating users and being informed about them throughout the development of HCD, or creating an arrangement where the service/system is developed iteratively at the users' site for an extended period of time design-in-use. In this differentiation, the various tools for user involvement, typically physical templates or digital forms, are seen as complementary resources to either techniques or methods (cf. Galabo et al. 2022).

These interrelations can be illustrated with examples taken from HCD, PD, and design-in-use-based participatory approaches. A good example is how contextual inquiry, contextual design, and HCD interrelate (Table 6.2). An approach such as HCD exists through being instantiated in several methodologies—one being contextual design. These, in turn, are composed of a combination of methods, such as contextual inquiry, that are, in turn, composed of several techniques, such as sequencing findings to sequence diagrams.

Now the most important issue to realize is that these interrelations feature varying specificity. Within methodologies, typically several different methods and techniques are used. Some of these methods are relatively specific to a methodology, such as contextual inquiry's specificity to contextual design. Looser connections can be found, for instance, in how future workshop methods feature in MUST (Bødger et al. 2004) (Table 6.3). A method can be used as part of different approaches if used somewhat differently, such as using service blueprinting cooperatively—not as an analyst's tool but in collaboration with user participants (cf. Muller and Druin 2012). Finally, there are methods that are used virtually across the board of engagements between developers

Table 6.2 An example of the relations between approach, methodology, method, and technique in human-centered design

Layer	*Example*	*Source*
Approach	Human-centered design	ISO 9241-210
Methodology	Contextual design	Beyer and Holzblatt (2008)
Method	Contextual inquiry	Holzblatt and Jones (2003)
Technique	Action sequence diagram	Anthropologists, sociologists, and systems analysts since second half of twentieth century

Table 6.3 An example of the relations between approach, methodology, method, and technique in participatory design

Layer	*Example*	*Source*
Approach	Participatory design	Robertson and Simonsen (2012)
Methodology	MUST	Bødker et al. (2004)
Method	Future workshops	Kensing and Madsen (1991)
Technique	Wall graphs of current work practices	Bødker et al. (2004)

Table 6.4 An example of the relations between approach, methodology, method, and technique in extended participatory design-in-use

Layer	*Example*	*Source*
Approach	Extended participatory design-in-use	Hartswood et al. (2002), Voss et al. (2009), Botero (2013)
Methodology	Co-realization	Hartswood et al. (2002), Voss et al. (2009)
Method	Placement of designer in the site where prototype is installed to continue design-in-use	Hartswood et al. (2002)
Technique	Dialogue about system-in-use at user's site	Researchers in ethnography and sociology of work, human-centered design, and PD

and users, such as some form of building mock-ups or qualitative interviewing (see Table 6.4 and Fig. 6.5).

The same goes for techniques. For instance, the specific way to combine watching and asking during fieldwork is specific to contextual inquiry, and usability think-aloud protocols are typically only used in HCD usability studies. In contrast, sequence diagrams of user action are commonly used in a range of approaches, be it HCD, collaborative design, or open design. Similarly, contacting participants is almost always taking place in one form

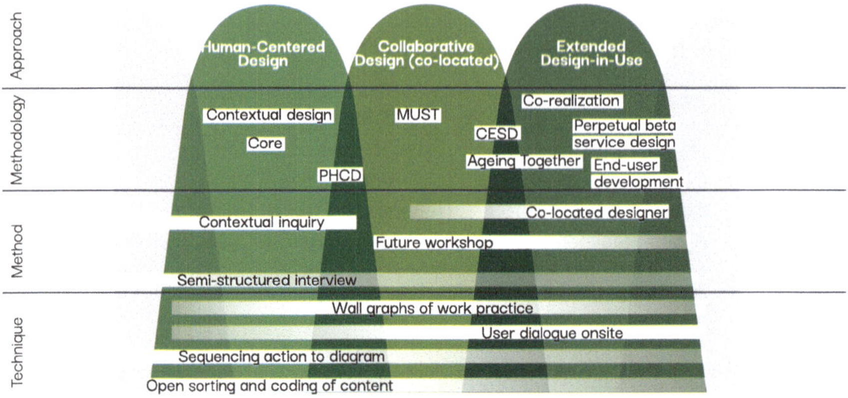

Fig. 6.5 A schematic portrayal of approach-methodology-method-technique relations

or another (part of necessary selection work that we explain in Chapters 2, 4, and 5). Even approaches that are particularly distinct in their orientation such as those emerging approaches in PD that have shifted design into sites of use for extended periods of time (e.g., Hartswood 2002; Voss 2009; Botero 2013) may feature techniques that are used for different aims and within different designer–user relations in other approaches, an example being dialogue sessions on a user's site about their practices and technologies-in-use (cf. Buur and Bagger 1999; Table 6.4).

From this, it follows that both the methods and techniques layers hold a considerable amount of category inclusion and "difficult-to-classify in and out memberships": not everything is part of everything, but many techniques and methods are included in several different kinds of methodologies and approaches, which in turn qualify how they are deployed. This, we argue, is at least one of the root causes for the difficulties both in mapping methods in the within-subfield typologies and in mapping the overarching features of the landscape. The mappings tend to get stuck on drawing and comparing items of different layers, extensions, and specificities, and particularly the layers of techniques and methods tend not to be specific nor define wider methodologies within which they are used.

Conceptual Starting Points for Distinguishing Approaches to User Involvement

Above we demonstrated that many problems in methods listings and mapping result primarily from attempts to map items that do not compare. These are particularly aggravated with respect to techniques and methods that can be deployed in conjunction with a variety of different kinds of methodologies. Somehow mapping them or including them, as is in mapping that also involves methodologies, may be a pursuit that is necessarily doomed to result in a considerable conceptual muddle.

But in contrast to mapping techniques or methods, we see hope in mapping methodologies being parts of approaches and differentiating between approaches in affinity to Iivari et al. (2000), if the interrelation between methodologies (instances) and approaches (categories) can be regarded as graded category memberships. Following prototype theory (Rosch 1973, 1975), a category is not seen as an Aristotelian additive sum of properties (bird = beak + feathers + hollow bones + born from an egg + has wings) or genealogically determined (see above the problems of cladism in domains without genes) but as consisting of elements with unequal status that impact its category membership, drawing from Wittgenstein's work on family resemblance. A graded membership in a category (a "family" of items) results in some items being more prototypical of the family than others, according to Rosch's (1973, 1975) definition of prototype. A sparrow is more prototypically a bird than an ostrich or penguin or archeopteryx, similar to how a chair is more prototypically a piece of furniture than a potted plant that retains a relatively more peripheral membership in the category of furniture. Similarly, in design participation, MUST (Bødker et al. 2004) is more prototypical of the class of collaborative design than "consumer idealized design" (Kaulio 1998), which would, nonetheless, have a lower grade membership within the class.

Prototype theory as an organizing principle for methodologies in design participation has several merits. First, attaching *prima facie* superiority or disciplinary lineage to different ways of conducting design participation has become an increasingly antiquated orientation over the last 20 years or so. Because of considerable crossovers and inventions of similar basic ways of working in multiple design-participation-related disciplines at the latest since the 2000s, disciplinary lineage no longer determines solidly where a methodology is located at (if it ever did). For instance, exploratory living labs (Leminen & Westerlund, 2012; Leminen 2015; Hyysalo and Hakkarainen 2014; Hesselgren et al., 2015; Hesselgren, 2019), design-in-use-based participatory design (Hartwood et al. 2002; Botero and Hyysalo 2013; Karasti et al. 2020), and collaborative perpetual beta industry strategies (Johnson 2013) land in similar extended real-life design-in-use engagement between designers and users, albeit from different origins. Just as well, living lab development pursued in a test-bed mode, user innovation toolkits, and open API ecosystems all depart from different disciplinary backgrounds yet factually end up in a user-involvement configuration that rests on a platform-based producer-to-many-user-inputs configuration. Hence, in terms of the instance–class relation, it has become sensible to begin to emphasize "resemblance" over "lineage" as in the case of living labs that factually land in two very different configurations by which designer–user relations are being organized (more on this below).

Second, resemblance is emphasized as the interrelations continue to change. Still in the early years of the 2010s, service design was mostly comprised of a set of methodologies and techniques that hold membership in HCD, collaborative design, and marketing, but it could then hardly be seen as a unique set of methodologies comprising a distinct approach at that point. Since then, the service design and related research has grown and expanded dramatically, but to the effect that user-involvement configurations within service design have multiplied rather than converged toward a clear approach (see the last section of this chapter).

Figure 6.5 provides one illustration of how approach-methodology-method-technique interrelations can be visualized. We illustrate this initially with three approaches and just one of the methodologies that instantiate it, and one method and technique used within each (see Tables 6.2, 6.3, and 6.4). Some dual memberships to approaches take place at the layer of methodologies already (the approach overlap areas), and this increases with methods and techniques as discussed above (white extension lines). We have added semi-structured interviewing as well as open sorting and coding to illustrate how some methods and techniques hold inclusion in a very wide variety of methodologies and thus approaches.

User-Involvement Configurations in Designing for, with, and by Users

In differentiating approaches, we examine the prototypical configuration that the methodologies within the approach hold in relating designers and users. The following elements are implied:

1. The broad rationale for seeking participation in a particular way
2. The roles and responsibilities designated for designers, users, and other actors
3. The arrangements by which solutions, domain knowledge, and need and preference information are carried between designers, users, and other actors
4. The arrangements by which design solutions are sought to be achieved.

By broad rationale for participation, we mean the rationale of why participants are *involved* in design. For instance, methodologies that reside within collaborative design share the rationale that users are experts of their own realities and can bring an understanding and creativity into the design process that complements that of designers through participating in design activities. While these methodologies differ widely in regards to why the incorporation of user realities is desirable for design—i.e., their normative rationale: improving profits, making work-affording systems, empowering the underprivileged, bringing more perspectives to design, democratizing technology development, deepening democracy (and so on)—the instrumental rationale for their *involvement in collocated sessions with designers* is broadly shared and remains distinctly different from, for example, approaches that rely on users doing the designing on their own.

The broad rationale ties to roles and responsibilities designated for actors. Continuing the above example, collaborative design holds designers as the experts of organizing and tooling the collaboration and refining designs developed together with other participants, whose main role and responsibility is to convey the needs and requirements of their contexts. Such a cast of roles is simply not held in open design communities or in investigation-based methodologies in HCD (ISO 9241-210) where designer roles and responsibilities for the design outcomes are considerably less and greater, respectively.

The arrangements for producing design solutions feature spatial and temporal and material differences. For instance, most collaborative design and UX and HCD methodologies trust that engagement between designers and users results in solutions that can be tested, iterated, and launched. In contrast, design-in-use methodologies such as co-realization (Hartswood et al. 2002; Voss et al. 2009) see work-affording systems only resulting from many iterations at the site of use, as their adequate shape and full potential is only then realized during this extended "play time" that allows for the co-evolution of the design and workplace practices (see Chapter 3).

Arrangements by which knowledge, ideas, and solutions are conveyed between the parties typically relate to the above and often make the prototypical methodological characteristics most visible. Crowdsourcing pursued with pen and paper is possible, but without an adequately designed platform, it results in such a taxing amount of manual labor that it can be carried out only in exceptional circumstances or as an addition to other methods (Hyysalo and Hyysalo 2018; cf. Chapter 5). Similarly, one can pursue design-in-use by inviting the users to periodic workshops, but it is far more effective to place the designer at the users' site or monitor the alterations that users make to the design, thus relying on different information transfer arrangements (Voss et al. 2009; Botero 2013; Johnson 2013; Karasti et al. 2020; Hesselgren 2019).

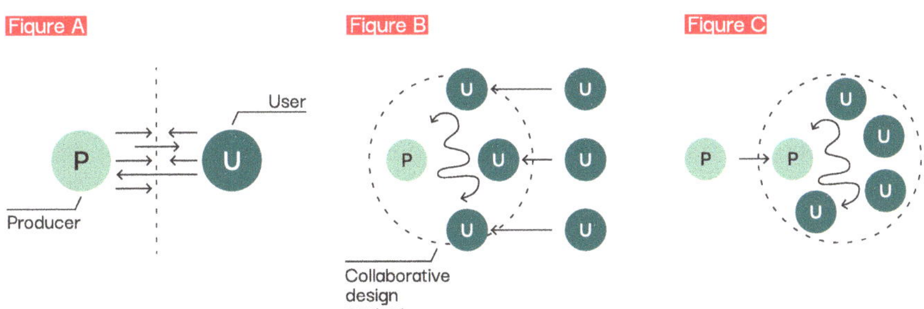

Fig. 6.6 Graphically depicting some of the key elements in user-involvement configurations

When discussing the configurations of different approaches, we further illustrate them with the help of a graphical mapping that originated in Heiskanen et al. (2010). The mapping depicts producers, users, information flow, and arrangements and shared participation context(s) (Fig. 6.6). In the notation, producer (p) stands for the party that seeks to produce the service or product under design be it a private company, a public organization or a non-governmental organization, and users (u) are the people implicated by it. In Fig. 6.6, the left image depicts no user involvement with an information transfer barrier (the dashed line) to be crossed. In the middle image, the producer (p) has changed the user-involvement configuration by creating a codesign context (dashed circle) to allow users to enter in design and to bring in insight from their background communities of practice (light gray circles around users). In the rightmost image, the producer has reversed the relation by entering the users' community of practice to design there, rendering the users' community of practice as a site for involving them in the design.

The participation configurations presented and compared here are depicted at the layer of approach, or "methodology family," and thus, they exclude considerations of, for example, particular project objectives or combinations of approaches used in the course of a particular project (e.g., Hyysalo and Hyysalo 2018; Botero et al. 2020; see Chapters 1–3). These are not pursued here to retain clarity over *approach differences*.[4] Let us now move to discussing approaches with these concepts and visualizations, departing from the approaches to collaborative design and its borderland area with user-community-based approaches.

Collaborative Design and Design-in-Use

Traditionally, collaborative design has meant taking users and other stakeholders in the design process as partners, often as experts of their own work or everyday practice (Bjerknes et al. 1987; Simonsen and Robertson 2013; Sleesvik-Visser et al. 2005). Interactions between designers and users have been predominantly collocated and mediated by tools that help users to design, such as easy-to-use work modeling, idea cards, mock-ups, and design games (Voss et al. 2009; Simonsen and Robertson 2013; Vaajakallio 2012). How the aims, responsibilities, design process, decision-making, control, and ownership of results are organized depends on the underlying rationale, be that the conflict perspective, participatory design, community PD, or simply codesign as means for getting desirable products and services onto the market (Bjerknes et al. 1987; Simonsen and Robertson 2013; Sanders and Stappers

[4] As such, this notation can be deployed at a project level and then allows adding more project-based specifics such as the effects of project objectives, changes in the user-involvement configuration over time, or the relation to the client organization or background communities of the involved users (see Heiskanen et al. 2010).

2008). Irrespective of these differences, the collaborative design methodologies rest primarily on synchronous and typically collocated user-involvement configuration (see Fig. 6.7a). Most projects have been limited to pre-launch stages of design, even as this is not a defining limitation. Because of this synchronicity and collocation emphasizing character, we choose to call this family of methodologies *collaborative design*, to which many prototypical methodologies can be found—a prototypical example being MUST, as we discussed earlier (Bødker et al. 2004). Synchronous collaborative design can be scaled up through a range of mechanisms, but the empirical fact remains that the participation has been limited to tens or hundreds of people, isolated exceptions aside. For examples of collaborative design approaches, see the workshops described in Chapters 2 and 5.

As we discussed in Chapter 3, the last two decades have seen the emergence of a range of methodologies that extend the traditional collaborative design into *design-in-use*. What is common to these methodologies is that they rest on a primarily *asynchronous* and typically *long-term* exchange between designers and users that continues *after* the initial product/service launch—a slowly paced, materially mediated dialogue that lasts for years (Johnson 2013). Such methodologies have emerged on one hand from within PD, prototypical of these being co-realization (Hartswood et al. 2002), meta-design (Fischer 2007), and aging together (Botero 2013), which all rest on design being continued at the user site for an extended period after the initial launch. On the other hand, just as important instances of design-in-use are exploratory projects conducted with living labs, where the bulk of

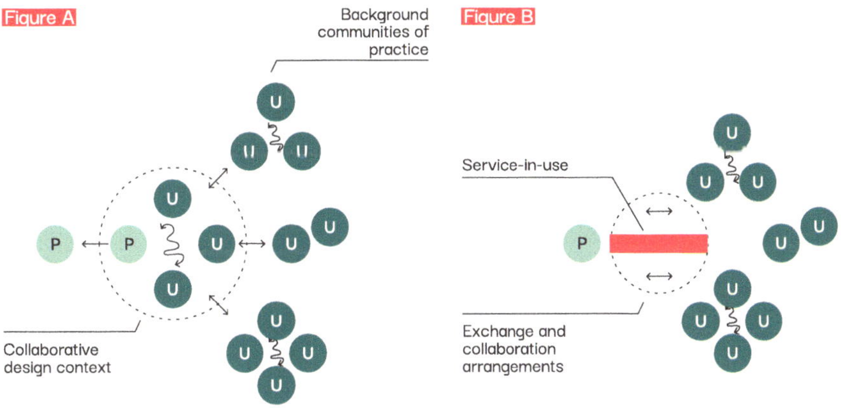

Fig. 6.7 **a** collaborative design, interactions between designers as representatives of the producer (P), and user (U) representatives, often from multiple background communities (light gray circles). **b** design-in-use, interactions between producers (P) and users (U) are predominantly asynchronous and temporally extended, mediated predominantly by evolving service/product and exchange and collaboration arrangements around it and its evolving usages

the technology design is carried out at the real user sites (Hillgren et al. 2011; Hyysalo and Hakkarainen 2014; Leminen 2015; Hesselgren 2019; Hesselgren et al. 2015). Similar gradual continued design-in-use projects have been pursued in industry contexts, often with design ethnography orientation at the background (Whalen and Bobrow 2011). Also, many methodologies within end-user development but outside PD are instances of design-in-use (Lieberman et al. 2006), as they build tools for programming competent users to self-serve in building and maintaining their systems and then evolutionarily redesign these tools as end-user projects take shape. However, the most numerous applications of design-in-use are found outside academia in minimum viable product (MVP) launch strategies of digital services (Johnson 2013).

In all design-in-use methodologies, the main mediating mechanism between designers and users is product-in-use (or service-in-use or infrastructure-in-use), on which both designers and users act upon and create at least new content if not form (see Fig. 6.7b). Due to the diverse origins and differences in specific projects, design-in-use features considerable variation in whether user involvement rests primarily on users' ideas or their solutions. When solutions are emphasized, users are more in the driver's seat in volunteering and suggesting new uses, new value points, and design solutions (Lieberman et al. 2006; Fischer 2007; cf. Kohtala et al. 2018; see Chapter 3). When resting on users' ideas, wishes, and needs, (co)designers take an active role in paying close attention to users' creative acts and how they articulate what is valuable for them in the design. This rests on the people's actions being ethnographically visible once they engage with product/service features, how they remix and complement the product or service, what new skills it creates, and so on (Hartswood et al. 2002; Johnson 2013; Kohtala et al. 2018; see Chapter 3). In terms of scale, extended design-in-use features a wide range, from design within a single workplace or community to hundreds of thousands of users contributing to evolving digital services in the course of their life cycle (cf. Botero 2013; Johnson 2013). For an example of design-in-use approaches, see Chapter 3 on the aging together approach (Fig. 6.8).

Coordinated User Design and Design Communities

There is a rising number of predominantly IT-based design participation methodologies that rely on a platform and a hub-spoke involvement configuration that does not require (and usually does not in practice include) interactions among the participants. Hence, no real community is formed among the participants or, if it is, it remains an additive element. We choose to call this *(host) coordinated user design* as it means creating technical and social arrangements by which users can independently provide solutions to company (or another host's) products or services (see Fig. 6.9, a). The prototypical

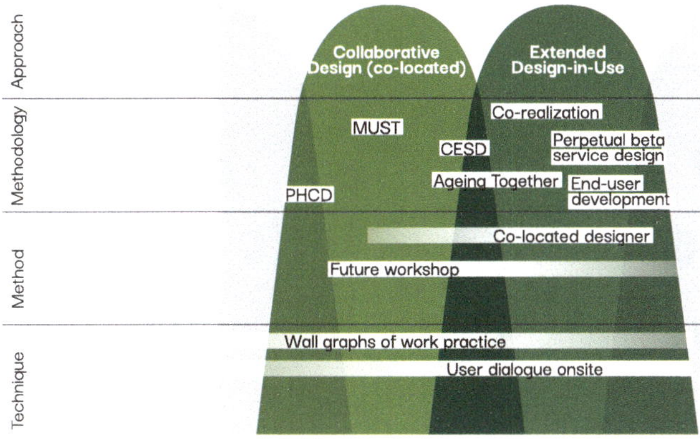

Fig. 6.8 Collaborative design and extended design-in-use approaches, with example variants and methodologies

methodologies in *coordinated user design* are open Application Programming Interface (API)-based ecologies, user innovation toolkits (von Hippel and Katz 2002), solution crowdsourcing, and the test-bed use of living labs (Leminen 2015; Williams et al. 2005). Less central but equally important members of this approach are customer contributed designs to mass-customization platforms and various "hosted activities" in the physical spaces such as museums, in which a designated location, conduct rules, and materials listings create a de facto platform into which users can design and bring their own events (Simon 2010).

The key marker of *coordinated user design* is a platform design that makes participants' solutions fit directly to host products or production processes with relatively small translation costs per user solution for the host. For instance, open API ecosystems allow user–designers to program (small) applications for a software platform and ensure they work if coded properly, which can result in very large added-value creation without massive investment from the producer, for example, resulting in millions of apps in Google's and Apple's app stores. Similarly, user-innovation toolkits provide designing users a finite module library of company production materials, which makes it easy for the producer to adopt and offer whatever users choose to design with them (von Hippel and Katz 2002). Coordinated user design is thus particularly apt when user-generated additional designs and variations can extend a more generic offering. Consequently, the scale of participation in platform *coordinated user design* can be very wide, and the hub-spoke configuration allows for upscaling the number of participants without excessive community management efforts and costs. The unfolding of participation is however predominantly limited to the making of one or a few small complete additions to the platform (e.g., apps, crowdsourcing solutions, or test-bed reports) and cut off from the overall evolution of the design endeavor, which is controlled

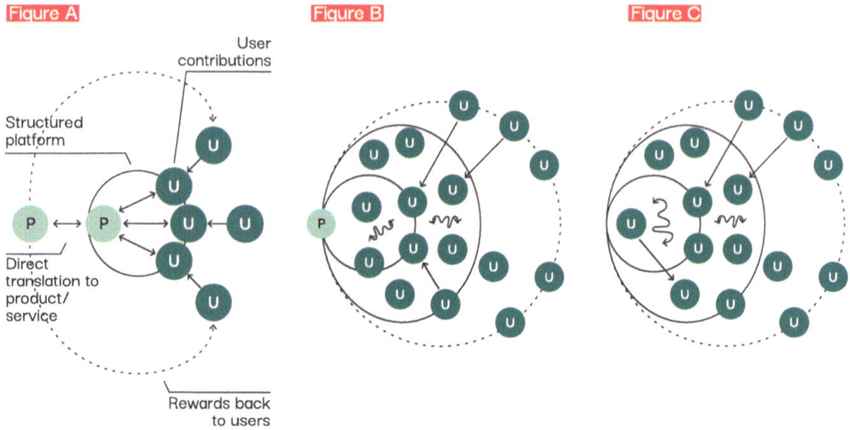

Fig. 6.9 **a** host-coordinated user design, where users (U) contribute solutions that directly fit producer's (P) product architecture through a structured platform. **b** and **c** user community-based design rests on user developers and outer circles of less intensively involved user (U) participants. In hybrid communities, a producer (p) is a central participant

by the host. A participant contributing to a test-bed or app ecosystem can, at most, lobby the producer regarding the overall development directions of a platform. This stands as an important contrast to the open user–designer communities we discuss below.

While the most well-known examples of host-coordinated user design are app ecosystems, the same configuration can be used effectively in more down-to-earth service and space development. An example of the same user-involvement configuration was the event organizing in Library 10 in downtown Helsinki, Finland. In the early 2000s, libraries started allowing and inviting user contributions, typically music gigs, reading events, and panel discussions by citizens and non-profit associations, to their premises. While these collaborations widened the library offerings and gave citizens avenues to organize their events, the drain on library resources was unsustainable, between 3–4 hours of staff time per event. To mitigate this, the staff used a red carpet to designate an area in the library space as the stage, created a box of wires, microphones and placed its content list in their website, created a code of conduct, and a simple reservation calendar. In effect, this created a (quite literal) platform onto which users could plan their events to be compatible and assess, for example, what materials they would need to bring or how to adjust their event to be compatible. The net result was that staff time per event reduced to 20 minutes, and event slots became fully booked by the citizens.

Moving to approaches that primarily rest on design efforts among peers, a relatively established division line runs between "independent" and "hybrid" innovation communities. The former includes (as prototypical cases) innovation by users, free- and open-source software communities (Ratto 2003; von Hippel 2001), open design communities of physical products based on digital sharing of blueprints (Balka et al. 2009; Abel et al. 2011), locality-based user-innovation communities (von Hippel 2016; Van Oost et al. 2009), and open content production communities such as Wikipedia/media (Benkler 2006; Ardati 2023). The enabling significance of IT sharing and the pooling of peer contributions is formidable in all of these, even as the locality-based communities and individual user innovators could manage without it (von Hippel 2016).

Hybrid innovation communities are found in open-source software projects hosted by a firm such as in Open Office (Freeman 2011) and design-intensive user groups and brand communities (Hyysalo, Jensen et al. 2016; Brandelli et al. 2008; Holmström 2004; Mozaffar 2016). Independent and hybrid user communities are worth keeping separate, as importantly different dynamics pertain to their community management, ownership, control, and the setting of development directions.

Common to all community-based user-involvement configurations is that participation among peers holds a range of orientations from inner-circle user developers to occasional design contributors and an outer circle of participants who test, report, translate, advocate, and use the community outputs (see Fig. 6.9, b and c). The temporal unfolding of participation tends to have "participation paths" formed through changing orientations and varying centrality in the community (e.g., Ratto 2003; Johnson et al. 2010; Freeman 2011; van Oost et al. 2009). The scale of participation in user community design can become very large, Linux and MySQL and Wikipedia as prime examples, but this requires setting up careful arrangements for managing the community and the contributions that it creates (Ratto 2003; Freeman 2011) (Fig. 6.10).

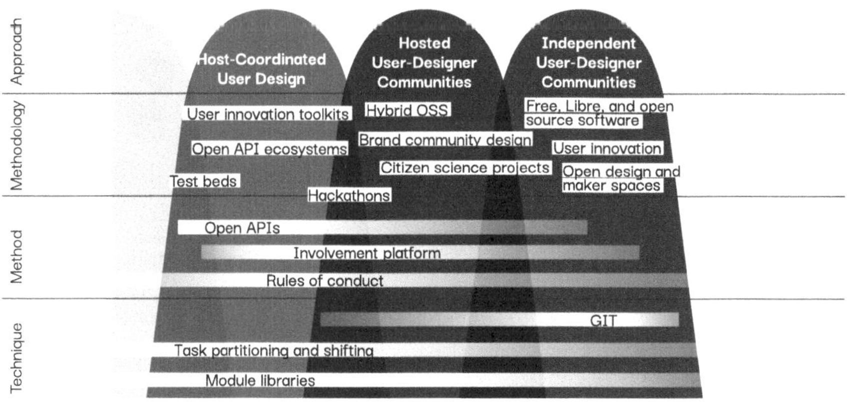

Fig. 6.10 User-design community-based approaches with example variants, methods, and techniques

Lighter User Involvement from Human-Centered Design to User Inspiration

Having now covered the approaches where parts or all design is shifted to participants, let us move to examine the approaches where some people are involved as informants and typically for relatively short periods, but involved nonetheless. In these approaches, it is rather the designers that seek to participate in communities and practices through some measure of participant involvement. Out of these approaches, the two most established ones rely on the investigation of users. Most of these methodologies are instances of HCD and its subset of usability engineering. HCD methodologies seek to give the needs, wants, and limitations of the end users of a product, service, or process extensive attention at each stage of the design process (ISO 9241-210). In all prototypical HCD methodologies, information about the contexts, requirements, and preferences of users is systematically collected, analyzed, stored, and moved to design solutions. The methods that comprise HCD methodologies have a wide range including ethnographic observation, interviews, surveys, artifact analysis, task analysis, role mapping, surveys, data logging, prototyping, and testing methods (see Fig. 6.11, a).

In the present millennium, HCD has largely given way to UX design, emphasizing the experiential, affective, meaningful, and valuable aspects of HCI and product ownership. It has evolved as a distinct set of methodologies, separate from HCD (ISO 9241-210). Instead of cognitive psychology and theories of action, UX draws from the psychology of experience and affects, as well as design disciplines (Hassenzahl and Tractinsky 2006; Law et al. 2009). Nonetheless, in practice this difference has been far less accentuated, and most practitioners consider UX to be an addition to HCD, as a kind of HCD+ (Law et al. 2009).

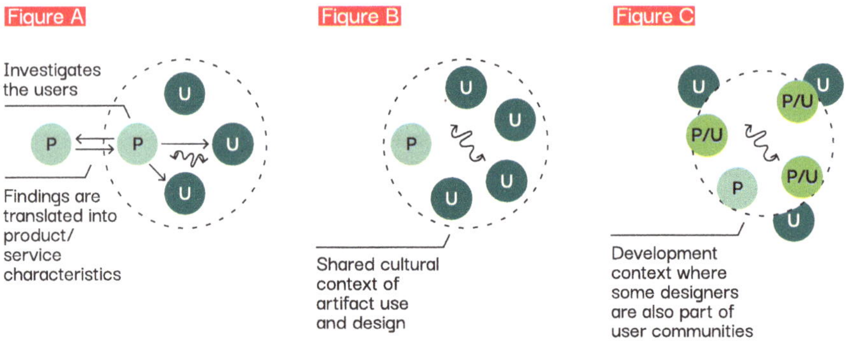

Fig. 6.11 In investigation-based approaches (**a**), designers study users in natural or laboratory settings and translate the insights into product/service characteristics. In user-inspiration approaches, designers are versed in the user domain of culturally mature products/services (**b**) or when developers hold dual membership in producer and user practices of new products (**c**)

It is important to realize that in-depth forms of marketing research such as those based on ethnography operate within the same basic configuration between developers and participants, and thus, the investigation-based approaches are not limited to those found within HCI (Hyysalo, Jensen et al. 2016, b). The same goes for universal design, ergonomics, and cognitive ergonomics insofar as they involve field studies or laboratory testing and do not just proceed with design guidelines and heuristics (which is an equally limiting condition to participant involvement in HCD and UX; not all the methods within these involve engagement with actual people either).

The final area in our mapping is conditions where designers hold a full membership in the use domain. This allows designers, insofar as they really are competent participants, to design in a responsible fashion based on mere *inspiration,* and any investigation, dialogue, or meeting being a secondary addition to this. We can discern two distinct types of such dual membership.

The first of these we call *designer immersion in use*. Designer immersion in use means that designers are competent practitioners in the specific user domain (such as a sport or a professional activity). This allows the designer to draw from a rich pool of understanding about who the other end users are, what requirements they have, what the contexts of use are, and so on. In effect, immersed designers can design for themselves and their peers and have easy access to gain whatever new information is needed, as well as easy access to checking and testing solutions (see Fig. 6.11, c). Prototypical instances of developer immersion can be found from well-documented industry cases (Kotro 2005; Heiskanen et al. 2010; Johnson 2013; Schweisfurth and Raasch 2015), where designers have been designing for their peers in media services and consumer goods. Such participation is typically limited in scale but often has extensive duration.

A well-known example of designer immersion in use is Suunto wrist-top computers. The project utilized marketing research and human-centered user research at its base. The whole product development team was active in the sports they were designing for, and drawing from their hobbyist knowledge in the sports made the team question the user and market research insights and shift focus from the high-end sports watch market to the “sports wannabe” market and identify design details that then mattered, resulting in a highly successful product line (Kotro 2005; Heiskanen et al. 2010).

The second form of designers being competent at use domain are cases of designing *culturally mature products and services* (Williams et al. 2005), such as tables, chairs, mugs, or artifact genres such as ATMs. In such cases, designers are already deeply familiar with the typical usages and use contexts of these artifacts through thousands of encounters with them (see Fig. 6.11, b). The designer’s participation in a shared domain can effectively act as the primary means of achieving responsible design. Additional insights and inspiration have been shown to still emerge from studying other people and their contexts and involving them in ideation, for example, through probes (Mattelmäki 2006). This said, it is important to stress that reliance on the "designer being user" works only insofar as designers really have sufficient participation in the user's domain. There is a rather grim history where this has not been the case, and a more participatory form of design would have been the socially responsible choice (e.g., Akrich 1992; Hyysalo 2006; Oudshoorn et al. 2004) (Fig. 6.12).

The demarcation line for when design participation is no longer taking place is porous and blurry. Manifold customer boards, citizen panels, and patient representative groups, for example, involve people particularly in the public sector but often with only a weak connection to design, planning, development, or sociotechnical transformation and rather limited on service delivery and its incremental adjustment (Bovaird 2007). Another demarcation line runs across when it is sensible to talk about *involvement* as low-level information provision is involved in traditional market and customer surveys, in customer feedback to companies, in informal exchanges between staff and customers, and so on. When such information gathering does not involve significant back-and-forth interchanges or other in-depth ways to engage with the actual people, it is best described as just gathering market information. This is pronounced in cases of data tracking, data analytics, and accumulation of customer information in CRM systems, which can be highly useful for gaining *insights* about people yet without involving them in any design-related activities (Johnson 2013). In contrast, some advanced customer research, for instance, may proceed through in-depth interviewing or ethnographic engagement and hence land in user inspiration, UX, or HCD approach clusters despite their quite different origin and orientation.

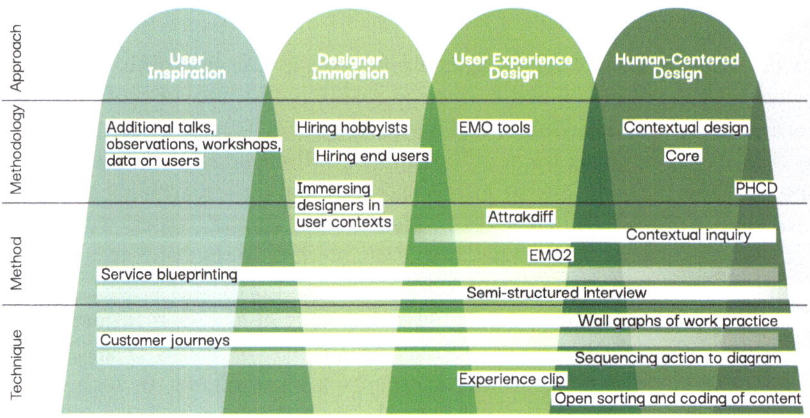

Fig. 6.12 Lighter user-involvement approaches with example variants, methods, and techniques

Wide Spanning Research/Practice Areas in the Mapping: Service Design, Participatory Design, and User Innovation Research

One of the problems we emphasized in the early parts of this chapter was that wide research and practice areas such as service design, participatory design, and user innovation research are regularly represented oddly in the various method and approach mappings. We are now conceptually equipped to show a key source of these misrepresentations, namely that these research and design practice areas are not confined to any one design participation approach but span several, typically owing to their historical evolution and influences from neighboring research and practice areas. In other words, such academic and practice areas are factually so diverse that they do not neatly "map" as entities.

Figure 6.13 elaborates the nine user-involvement configurations we have discussed in this chapter and depicts their relation to three research areas that regularly cause confusion regarding how they stand with respect to user involvement. The first of these traditions is user-innovation research (von Hippel 2005; von Hippel 2016). The thrust of this research and practice community is in capacitating users to innovate and turning this into a company, non-governmental organization (NGO), or public-sector benefit. Examining its major communications outlet, Open and User Innovation Conference 2004–2022, the research spans from independent user innovators and independent user-innovation communities to ways by which companies can work with hosted user communities and onwards to various user-innovation toolkits and crowdsourcing platforms and contests (for example works see, von Hippel 2001; von Hippel and Katz 2002; Balka et al. 2009; von Hippel 2016). But this research community does not pursue user-involvement configurations in which users do not perform independent design or innovation tasks (Fig. 6.13).

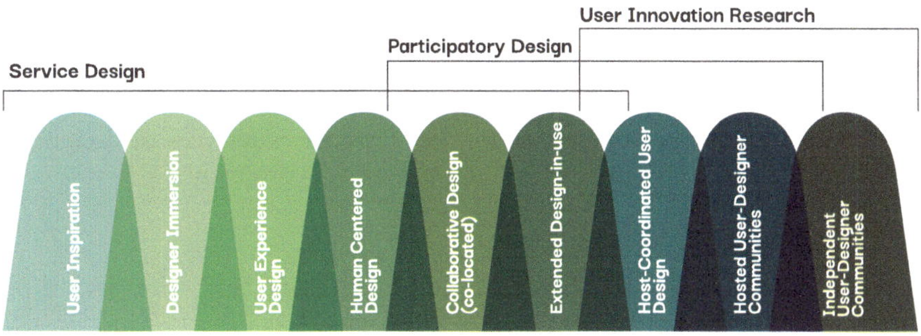

Fig. 6.13 Mapping the range of approaches differentiated by their involvement configurations, the borderline of user involvement, and the spans of three large research/practice areas

Participatory design, as described above, has fostered synchronous collocated collaboration since the 1970s (Bjerkness 1987; Bødker et al. 2004) and design-in-use approaches since the beginning of the 2000s as well as liaisons with independent user–designer communities and initiatives to set up hosted user–designer communities and exploratory living labs (e.g., Hillgren et al. 2011; Botero and Saad-Sulonen 2010; Huybrechts et al. 2025). Some of the public space projects feature small independent user designs that would fall within host-coordinated user design (e.g., Simon 2010). Participatory design thus engages in five distinct types of user-involvement configurations, although it should be noted that in comparison to user-innovation research, it is more hands-on when it engages with communities and typically with less design and innovation intensive communities (e.g., Karasti et al. 2010; Hillgren et al. 2011; Hyubrechts et al., 2025) (Fig. 6.13).

In turn, service design has a wide application area that includes several types of design and research that feature very different user-involvement configurations. At one end, service design agencies make incremental space, service encounter, and digital app designs in which user involvement remains inspirational or absent (cf. Patricio and Fisk 2011). In more encompassing service design projects, users are investigated and designs are tested in a user-involvement configuration typical to UX and HCD. Many, particularly Scandinavian, service design projects feature workshops with users and stakeholders as the prime user insight method (e.g., Hyvärinen and Mattelmäki 2015), landing in a collaborative design configuration. Finally, many digital service design projects, particularly when pursued through agile development and early releases, proceed with design-in-use (e.g., Johnson 2013). In all, due to its deployment in spatial, physical, and digital services—all with their distinct development processes—service design presents a prime example of research and practice that is so diffuse regarding its design participation configurations that any attempt to isolate commonalities across a small set of prototypical methodology is unlikely to ever succeed. Yet the human and customer orientation are arguably the most defining characteristics of service design regardless of the way by which it is pursued, and better understanding of the different roles and mechanisms by which participation is factually deployed within service design is key to orienting service designers to different facets of their trade. Table 6.5 further condenses the key differences in roles of designers, participants, and user knowledge in different user-involvement.

Chapter Conclusions

Mapping differences between different ways to carry out design participation is wrought with difficulties. In this chapter, we have introduced the common quandaries and then proposed that, nonetheless, some zoomed out differentiations can be made, premised on predominant design participation configurations and graded membership of methodologies within approaches. Such mapping reveals important, and routinely neglected, differences in scale

Table 6.5 Designer and user roles in different user-involvement configurations

Designer-user configuration	*Designer role*	*User role*	*Role of user knowledge*	*Information arrangements*	*Solution outcome*
User inspiration	Creative professional	Muse	Enhancing designer vision and imagination	Designer's own experience complemented by involving or studying users	Insights that designers turn to concepts and into production
Designer immersion	Design and user expert	Peer, informant	Grounding for design	Immersion in both user and producer practices and contexts	Insights that designers turn to concepts and into production
Investigating use and users (HCD, UX)	Investigator, protector, and developer of users' life	Subject of investigation	Grounding for design and evaluation	Observation, interviewing, task analysis, UX testing, etc.	Insights that designers turn to concepts and into production
Collaborative design	Organizer, facilitator, design realization	Design participant, expert of one's life	Users' knowledge to design	Co-located workshops, site visits, etc.	Shared design decisions that designers turn to concepts and into production
Extended design-in-use	Organizer, facilitator, design realization	Design participant, code-signer	Users' solutions and knowledge responded to	Design-in-use and information passed about it	Ideas, requests, and solutions iterated into evolving system or service
Coordinated user design	Platform designer	Additive designer	Starting point for solutions	Platform or toolkit that fits producer process	Independent design additions
User communities (hybrid, indie)	Facilitator, supporter, peer designer	Designer, realizer participant	Starting point	Passed between participants	Evolving product or system

and temporal dynamics in participation. This provides a discipline-agnostic way to highlight important differences in participation configurations that have sprouted during the last 30 years. We believe this is an antidote to common (mis)perceptions among practitioners, academics, and students of what is involved in different ways of pursuing design participation.

In practical projects—both our own and others—the purposeful hybridization, mixing, and sequential deployment of different engagement registers is common, as discussed across the previous chapters. Yet our experience is that the differentiations pursued in this chapter are most needed exactly when a shift in designer–user relations takes place in the course of the project. Without clarifying devices, project participants may fail to recognize that they need new sensitivity or a new kind of means or participant roles or timeframe for unfolding the work.

The typology is not without its own limitations. The prototype and class constructions cannot be as evenly empirically derived as one might wish. Some of the old and academically rooted ones are well-recognized labels with clear definitions as well as many methodologies within the cluster—for instance in UX, collaborative design, and open design communities. In contrast, some no-less-important practitioner's commonly used approaches, such as designer immersion-in-use, are far less academically defined. Some categories are emerging and hold overlapping "native" terminology, currently particularly in the approaches to extending design-in-use and (host) coordinated user design. Our typology downplays some common real-life forms of developer–user engagements, such as innofusion, where technology continues to be innovated at the sites of use but without planned arrangements and often to somewhat dysfunctional processes. In the present typology, this would classify as "non-planned design-in-use" (Fleck 1988; Williams et al. 2005).

We wish to underscore the importance of "zoomed-in" processes in how developer–user configurations factually play out, as detailed across Chapters 2–5. There are key contextual differences that matter in how user-involvement approaches become deployed, for example, due to different client relationships in between designers and eventual users and whether the design participation is foremost pursued because of corporate profit vs. citizen empowerment—the normative and instrumental rationale for participation cannot always be so neatly kept separate (Jensen and Petersen 2016; Botero et al. 2020; Palmås and Busch 2023).

Finally, rather than a carved-in-stone taxonomy, new ways to user involvement continue to emerge and stand out from the established user-involvement configurations. Helping designers, students, and fellow academics to recognize what is novel and how exactly it is novel is one of the facts the analytics provided in this chapter hopefully help students, practitioners, and academics to do.

CHAPTER 7

Scoping and Planning Design Participation

Introduction: Toward an Analytical Approach to Planning Design Participation

This penultimate chapter of the book shifts focus to elaborate on an analytical approach to gauging if and what design participation actions could be (best) suited to a particular initiative toward social and/or environmental change. The book overall has advanced an argument that there is no one right way to pursue design participation but also that this by no means entails that "anything goes." There are many ways to go astray and waste resources in pursuing it. Design participation must thus be adjusted to the contexts, aims, resources, and opportunities that are present. How to approximate this framing work challenge—either heuristically or in a more rigorous fashion—is the focus of this chapter.

The contextual approximation pursued in this chapter differs somewhat from how contextual considerations in human-centered and participatory methods have proceeded (cf. Beyer and Holtzblat 1999; Slesvik-Visser et al. 2005; Bødker et al. 2004; Sanders and Stappers 2012). Most contextual analysis has focused on clarifying what is the sociotechnical context for a product or service to be designed, usually with an already selected method or toolbox (cf. Beyer and Holtzblat 1999; Checkland 1984; Slesvik-Visser et al. 2005; Bødker et al. 2004; Sanders and Stappers 2012). An inquiry into the customer/user context has been carried out to determine specifics of the context to better tailor the methods or approach chosen, typically by the way of some measure of qualitative field research such as ethnographic observation and interviewing. Such contextual approximation types are a pivotally important phase of the framing work in service and product design projects (Slesvik-Visser et al. 2005; Szymanski and Whalen, 2011).

S. Hyysalo, *Design Participation*,
https://doi.org/10.1007/978-3-031-95437-5_7

But what we embark on in this chapter is one step *prior* to this kind of more customary contextual approximation—*before* choosing what may be the approach to be taken to begin with. This is because the generation of insight about the context may take a fundamentally different character if, for instance, the chosen approach is sparking a community design effort (with participants who hold rich and multisided context understanding) or a designer/ researcher-driven endeavor where most context understanding needs to be first attained.

This chapter outlines a set of considerations for this early approximation process, which also tends to give a good start to the more traditional and detailed contextual study of the targeted contexts of design participation. We do so by using concepts introduced over the course of the previous chapters from design research, science and technology studies, innovation studies, and transition studies.

In a small and relatively quick project, such contextual approximation is necessarily quick and heuristic—just a few hours of considering the contextual conditions affecting the endeavor—in a sense making a considered guesstimate about where the contextual "sliders" introduced below land and what this entails.

In more serious and "heavy-duty" projects, the approximation necessarily needs to be more analytic and drawn out (for similar consideration on single-method and technique deployment, see Slesvik-Visser et al. 2005). And if nothing else, the reason for seeking to approximate the matters bearing on design participation is to work out whether one can sensibly proceed in any other way but in heavy-duty fashion, or, at that, whether it makes sense to pursue design participation at all (indicating for instance that some other action toward environmental and social change is clearly more effective).[1]

[1] The constructivist stance on taking design participation to be an emergent relational outcome—of participating subjects, objects, and formats of participation (Chilvers and Kearnes 2020) or of matters, forms, and produced participations (Hyysalo and Jensen et al. 2016) or assembled actants (Marres 2012; Kelty 2020)—results in an interesting conundrum regarding its implications for practice. For an analyst, a constructivist and relational approach to participation allows for postponing any take on the constituents of design participation situation(s) and setting(s), to be settled only during the empirical analysis, such as asserting which actors, sociotechnical structures, infrastructures, and politics are factually present and the most important ones and not just assumed based on some theory or on investigators' assumptions. This empiricized ontology (Marres 2012, following Mol 2002) further implies that the empirical or normative bearing of any entity in the assemblages that form participation cannot be pre-fixed but depend on its relations to the other actants at play (Latour 2005). This includes what might be the exact issue and matters and consequently who exactly might be the actors that would form the insider and implicated actor groups (Clarke, 2005; Marres, 2012). For an analyst, this gives much needed freedom to attend to what really happens in design participation. Yet, for a practitioner wishing to engage others in a cause or an issue, such postponement just does not quite cut it. The practitioner cannot only wait and be reflexive about their decisions and indecisions (cf. Chilvers and Kearns 2020) but *must proceed* to a ballpark estimation of how to sensibly get started and carry out the design participation work in order to accomplish it. This can feature reflexivity in action, and can be further reflected on (Schön, 1982). And surely,

Not pretending to provide an exhaustive list of matters to be considered and insisting on the continuous translation between design participation choices, we next walk through three cases in which we currently use design participation in energy and climate transitions and which are all instances of more commonly encountered situations where design participation is pursued. Using just one domain area hopefully helps grasp what difference the assessments make for potential design participation avenues, while the many cases presented in previous chapters on healthcare and cultural sectors should help careful readers in translating the insights to their own domains.

Clarifying the Matters and Issues for Design Participation: Case District Heating Mismatch

Let us begin with a case that is of one type for which design participation is commonly pursued: two or more actor groups having the potential for a more mutually beneficial relationship that just has not materialized. There appears to be a common benefit to be had, but tapping into it remains somehow blocked, and it remains blurry why so. Our experience is that this kind of mismatch-cases abound in society, for example between city planning and neighborhood needs (e.g., Vertiz 2024); between public health offerings and residents' needs and resources (Hyvärinen and Mattelmäki 2016); between owners of unused property and potential new kinds of tenant groups (Hernberg 2022; Hernberg and Hyysalo 2024); between city administration and urban activists (Hyysalo et al., 2023; Pirinen et al. 2022); and between producers and users in new circular business models (Marttila et al. 2023). What is by and large shared across such cases is that there is some ignorance between parties and potentially some deeper dynamics hidden to the outside eye that keep the parties from engaging with each other in more mutually beneficial ways. And quite commonly, it occurs either to some of the parties involved or to an interested outsider that design participation would be worth a try to solve the riddles involved. But how to best do that is quite context specific.

there should be iteration and gradual clarification of the context, but if that is *the way* to proceed, it is already a commitment to an iterative and incremental way of organizing the design process—one well suited for some endeavors but less so for others. Thus, and somewhat ironically, the relational ontology that seeks to avoid proceduralizing and instrumentalizing participation and emphasizing its situated nature results in a *greater*, not lesser, need to begin to clarify of the key contextual constituents of design participation from the onset of whatever design participation action is to ensue. To hammer this point home: if the adequate formats of participation are not taken as given, let alone as universal, to be chosen and deployed in a pre-determined fashion in all settings alike with minor adjustments, then the setting and situation of participation, including its matters, actors, and structurings, must become clarified to a considerable length in order to adequately choose and adjust or design the format(s). Luckily the proverb of an egg-laying chicken does not quite hold here—an understanding of the situation can be co-constructed during the evolution of design participation even as the co-construction must first start—and it is the early assessment in this approximation process of the relevant contextual parameters that the present chapter hopes to aid.

The example case with which we elaborate this design participation situation concerns the use of **ambient and waste heat reserves in district heating**. The background is the advancing energy transition during which the old paradigm of burning fossil fuels in centralized district heating plants to generate heat (and often power as a side commodity) has become untenable due to carbon emissions and emissions prices. Shifting to non-fossil district heat generation can be done with large-scale heat pumps and high efficiency large electric boilers, which would be all the more efficient and cost-competent if the source heat reserve they use is of a higher temperature than otherwise available. Such excess and ambient heat is, in turn, available from, for instance, sewage plants, data centers, industrial plants, and just as well from supermarket freezers and refrigerators and residential buildings that may have cooling needs or produce energy in excess for their own needs. In turn, for these excess heat providers, the uptake of excess heat is often a cost saving as it relieves their cooling needs or at least is a resource they could not otherwise really use. The excess heat providers and district heat companies may thus have a potential win–win, and many successful installations already exist. Yet, at issue here is the fact that despite the local availability, few of the 250 district-heated towns and cities in Finland have yet tapped into this resource, and there appears to be virtually no public debate about the topic either (Auvinen 2024).

To gauge if and what kind of design participation could be suited, we can start by approximating the **issue characteristics**, **structuring conditions,** and **actor group** dynamics we need to consider.

The very first item to clarify is what are the "matters" and "issues" that the design participation addresses and how they relate to change domain actors within that. Sometimes there are socially salient issues that have come to (mutually) define the public(s) involved (Marres 2012). An example of such a situation would be climate change and its relation to climate action by various activist groups such as Extinction Rebellion or ecological show homes, public media, the Intergovernmental Panel on Climate Change (IPCC), climate negotiations, and government responses in different countries and political coalitions. An issue defines the affected people to whom the issue is *relevant*, who in turn define the issue characteristics in participating in the public, and through intertwining the issue in the "going concerns" of institutions and organizations (Hughes 1993; Blumer 1969; Becker, 1982; 2008). What are the matters and issues involved is thus among the first things to consider. Here, increasing the use of ambient heat reserves in district heating remains *weakly issuefied* in public discourse in Finland, as it is recognized and articulated only in a relatively small circle of social media groups such as the "new energy policy" discussion group. At the same time, it is an important matter for a small circle of experts and enthusiasts who in principle can also affect it. They do recognize it as important and carry some public and private debates about it, even though this seldom carries to to mass media.

Related to issuefication is the *familiarity of the matters involved* to potential participants. At one extreme are matters that are culturally familiar and so rehearsed that people recognize them and practically "everybody can describe well," that for example diets and everyday eating habits and the potential of rendering them more low carbon might be. Whoever the participants are, they would at least have informed opinions on the matter based on their own experiences and media discourses. At the other end of spectrum are the poorly elaborated, emergent, fuzzy, and/or complex matters that even experts struggle to understand—a good example being the case in Chapter 2 about what hacking and making practices may be comprised of in 10 years' time. As we learned in Chapter 2, these kinds of settings set great demands for who can meaningfully participate and how their participation and decision-making would need to be supported.

The use of heat reserves in district heating falls in between these extremes but clearly toward the poorly defined and expertise-requiring end. Within the limited circle of energy system professionals, the potentials and barriers to using ambient heat reserves in different district heating companies is subject to debate, opinion, and unclarity, often owing to high local variation in the amount and kind of heat reserves and specifics of the district heating network in question.

What about the **structuring conditions** bearing upon ambient heat utilization in district heating? A good way to start is to consider which structuring conditions are central to the focal issue or design matter and start listing them. If one wishes to proceed visually, one visual tool we prefer to use in the early assessment is the "situational matrix" by Adele Clarke (2005) (Fig. 7.1). It is a way of depicting elements that constitute the situation of action, without yet *theoretically* pre-assuming what kinds of interrelations between situational constituents exist, as these can then be traced for the matter at hand *empirically*.[2]

[2] A situational matrix differs from the common approach to thinking about context through levels, such as micro, meso, ekso, and macro level (or system) in either vertically distinct spheres (e.g., Bødger et al. 2017) or circles in which smaller units are enclosed by larger entities (Strauss 1993). In such theoretically ordered levels, understanding individual actions and interpersonal interactions are typically taken to form the micro level, which is then surrounded by, for instance, work group, community, and organizational levels, and these are again enveloped in an industry sector, a nation-state, and so on. As remarked by Peter Hall (quoted in Clarke 2005, p. 70), "[T]he imagery of… concentric circles, while perhaps a heuristic device, conveys an erroneous vision of social topography, one that I would rather leave to empirical examination" with a pre-cast social topography of types of actors and actions that are pertinent in each level (e.g., "organizational"- and "institutional"-level actors). It remains odd to think that, for instance, "economic" or "political" or "climate change" would reside only at a greater remove than, say, "action" and in need of somehow being translated into it. They are rather right there, even if they are also elsewhere, and clearly, they are not uniformly present in all sites and settings, let alone continuously (Hyysalo 2010). Thus, practically, the concentric circles topology would imply an analysis of all the layers and their effects, which would mean a massive enterprise that requires different types and granularities of data even when dealing with relatively straightforward and small events (Strauss 1993, p. 64), further complicated by epistemic issues regarding the compatibility of the mechanisms by which spheres affect another. Thus, an alternative way for thinking about the contextual relations we prefer is

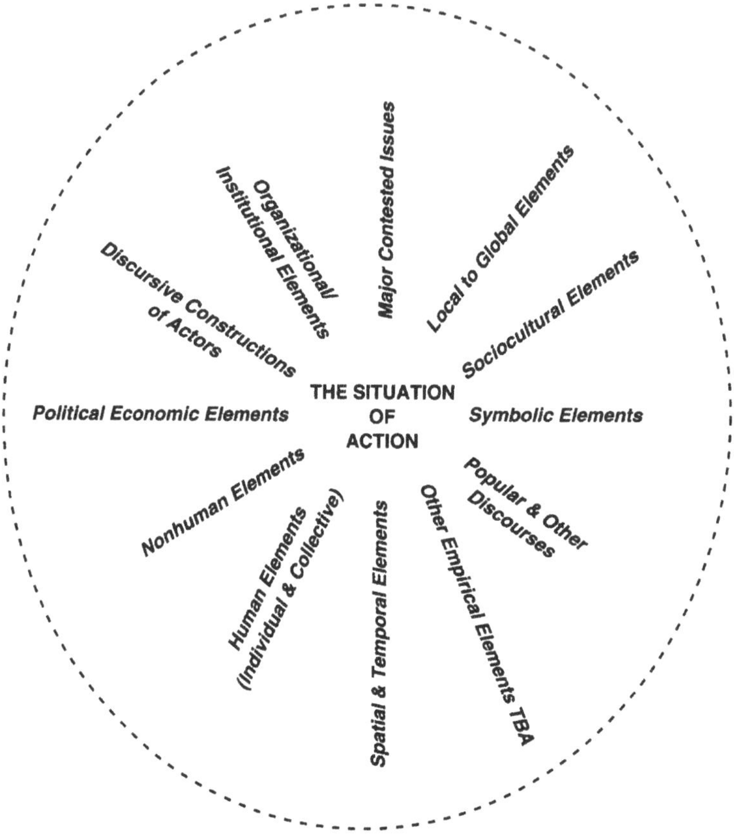

Fig. 7.1 Clarke's situational matrix (Clarke 2005, p. 73). Reprinted with Permission

Surely, the theories we have encountered over the course of this book help *sensitize* us to assess the presence and, often also likely, bearing of many sociotechnical entities that tend to feature importantly in many design participation situations. The idea then is that, through familiarizing ourselves into the context, we can gradually nuance our understanding of the situation and "populate" our view of the matters, issues, and ecologies of actors involved.

to admit that the context of a situation (e.g., sociotechnical structures, economy, politics, climate change, negotiations, activist campaigns) is that which is being woven together into a definition of the situation for the actors as well as for the researcher. Reality and situations are constructed in situ (Lave, 1993), but we are never alone (or) at its sole construction site (Latour 1987). The question of context gets thus transformed into asking how these conditions appear—making themselves taken as consequential—inside the empirical situation under examination (Clarke 2005, pp. 71–72). This is thus a more empiricist view of context that we prefer, while adopting that previous research gives us *sensitizing* insights about what contextual conditions we would need to gauge regarding the situation of action.

For our district heating case, an early situational matrix could look like Fig. 7.2, depicting at least those conditions and actor groups that are saliently present but not yet assigning interrelations or relative importance among them.

To then start taking closer empirical bearings on the situation, we can start approximating the presence and gravity of some of the common sociotechnical structuring conditions we encountered in Chapter 4. There we discussed path dependencies, infrastructural lock-ins, and interlinkages and how these may result in high obduracy and even "sociotechnical regimes," highly stable

Fig. 7.2 Tentative situational matrix for ambient heat utilization in district heating

sociotechnical networks that often also feature wide societal and infrastructural "reach" (David 1985; Geels 2002; Truffer and Füenschilling 2014). We there underscored that if the domain and issue at hand is constituted against such intensive sociotechnical structures, no alternative design solutions and efforts at design participation are likely to have much effect. An example was how electric and shared mobility withered for decades to the effect that viable alternative design concepts never made it beyond the car shows or early demonstrations before becoming canceled (Hoogma et al. 2001).

Similarly, strong structuring may ensue from conflictual political, interest-group, and value-laden positions. In a highly politicized situation, an attempt to initiate a change initiative is easily subjected to targeted reactions for or against. In today's identity politics and affective polarization-ridden democracies, this may happen regardless of what the original intent for the social change initiative was.

But the whole world is not made out of sociotechnical regimes or political tug-of-war. Even when such regimes and interest conflicts are present in some domain, there may be aspects that are of relatively low importance from the regime actors' point of view. At the other extreme, there is thus a relative lack of structural resistance to at least small-scale and initial change processes. It could also be that the issue is located within the sphere of private action or private property to an effect that any external actor would find it hard to meddle with it. Such relative lack of structural conditions might also have to do with the issue characteristics in that the issue is so nascent that actors do not yet hold positions toward it or have not yet worked out how they may, could, or should be affected by it (Fig. 7.3).

The tentative answers to the structuring of the situation of action begin to direct attention to the question of what the key actors and actor groups are that exert stability or change in the matters and issues at hand. Instead of just defining the actors, actor groups, or publics with respect to the technology, matter, or issue at hand (cf. Bijker 1995), one would want to make a further move and list/map what these actors primarily do—in other words, what their prime concerns are. This is because, for whatever the issue at hand is, it is likely just one commitment for most actors implicated by it, and potentially not even their most important commitment (Hyysalo 2021; Hyysalo et al. 2022). Alternative ways to phrase this are to examine what communities of practice (Lave and Wenger 1991; Wenger 1998), social worlds (Strauss 1978; Becker 1982; Clarke and Star 2003), or activity systems (Engeström 1987, 2000) are involved, and how. Clarifying this helps to get a better sense of what the interests of different actors are regarding the matters at hand.

Furthermore, one can proceed to map out the *arenas* in which the actors are involved and which they reciprocally come to define through their presence. The importance of paying attention also to the arenas in which the sociotechnical issue is acted on is twofold. First, many arenas relevant for design participation impose their own conventions and rules for interaction as well as limit which issues and matters can be brought onto them. For

Regulation in the domain

LOW HIGH

In high-regulation contexts most change has to go via regulatory change.

Moderate to high: District heating forms a natural monopoly that must adhere to regulation. Within it, the district heating companies do have considerable discretion as to how they organize and charge the heat provision.

Depth of path dependencies

LOW HIGH

Path dependencies are difficult to reverse; they may require shielding and nurturing measures.

High: District heating plants and networks have been built for centralized fossil fuel-based heat generation. Even as the fuel choice has been destabilized and often shifted to wood, pellets or waste, the rest of the structures remain.

Sunk investments

LOW HIGH

Actors standing to lose much from the change likely engage in public or covert resistance.

Moderate to high: Investments in power plants and heating pipes have long payback times, renewal cycles depend. Investments in plants and buildings inducing excess heat also long.

Prestige and power loss

LOW HIGH

Powerful actors may resist change if it means losing power over their domain of action.

Moderate: District heating is mostly municipality owned, local sourcing of wood and peat locally endorsed.

Infrastructural interde-pendencies and lock-in

LOW HIGH

Infrastructural lock-ins are difficult and costly to reverse.

High: Power plants, heating pipes and heat exchangers in place - lowering district heating water temperature requires replacements. Commitment to excess heat provider and provision has to be long term.

Fig. 7.3 Assessing the presence of structural sociotechnical conditions, ambient heat reserves in district heating as an example

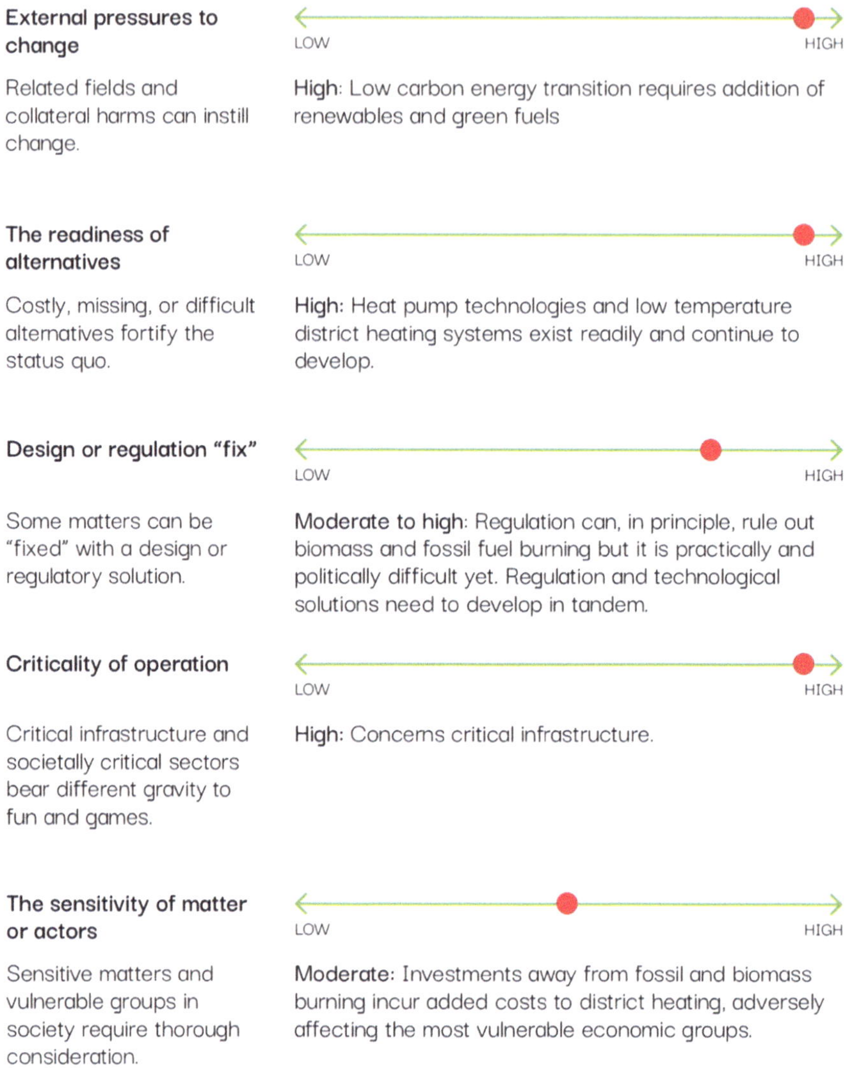

Fig. 7.3 (continued)

example, participating in a regional planning arena is effective only if one follows the procedures and timespans of a given urban planning endeavor and often also the conventions and notations used. Second, some of the actors or actor groups may be (and usually are) connected not in just one arena but in several, thus forming an interlinked meshwork of actors and arenas. In mature industrial fields, this may cohere into a "business ecosystem" or wider sectoral "technological innovation system" (Markard 2018), but in more emergent and nascent issues and contexts typical to design participation, the actors involved may not yet have such coordinated interrelations, and its thus better to characterize them and their interrelations as an "ecology of actors" (Hyysalo 2021; Hyysalo and Juntunen 2024).

With these basic parameters in place, one can proceed with the first approximation as to what the actor positions and coalitions are and how fixed they may be and what agendas different actors may have regarding the issue at hand. Furthermore, as engagement with the change domain accumulates, this mapping provides a sense of how the different actor agendas "carry through" to practice, for example, where points for change/resistance/agreement can be found. Such mapping also provides what the issues are that are bundled together by different groups and actors in different arenas and even between arenas (Hyysalo 2010; Hyysalo et al. 2022).

Such tentative mapping of the *ecology of actors and arenas* is important for next clarifying different actors' and actor groups' *willingness* and *capacities to engage* with the issue at hand. Clearly, different actor groups and actors within them hold different interests and willingness to engage in a given issue, and, while the willingness can be coaxed and supported by making the participation easier, process-wise more rewarding, and more impactful, a ballpark estimation is good to pursue from the onset. The same goes for different actors' capacities to engage. Some issues require relatively everyday capacities of action, such as changing one's own behavior, refurbishing one's home, voicing one's opinion in elections, or organizing a protest event, a march, or boycott campaign. But as we learned in Chapter 2, there are issues that require more design-intensive capacities to be acted on effectively, for instance, contributing content and informed alternatives to planning processes and regional strategies. When such specific competencies become requisite for affecting the issue, the default case is that some actor groups will hold much higher expertise levels, resources, and power positions than others. The implications for participation are profound. As we saw in Chapter 2, meaningful participation may require high domain knowledge and design ability, effectively questioning the meaningfulness of representative participation or efforts at bringing marginalized groups into design and decision-making. As we saw in Chapter 4, the expertise and abilities can also be highly partial and asymmetric, which warrants participative actions premised on complementary competencies such as in diverse actor transition arenas.

In the district heating case, an obvious key actor group is district heating companies that could adopt excess and ambient heat in their heat provision either directly or through refining the heat to a higher-grade heat via heat pumps. As noted, their size, ownership, technology base, and future orientation vary greatly. The second key actor group is even more diverse, and this is the excess heat suppliers, which for one reason or another could offer a heat resource to district heating production. These actors include industrial plants, data centers, commercial buildings, public-sector property, and landowners and even residential associations. These actors are primarily linked to a range of quite different sociotechnically constituted arenas than those providing a heat reserve. Most people in these two focal actor groups feature higher education backgrounds and are competent in elaborating their interests and solutions.

In addition to these two focal actor groups, regulatory bodies are central actors. These include market and energy market regulation actors, land use planning and permitting authorities, and actors advancing climate-change-related environmental regulations nationally and EU-wide. Also, towns and cities and their politically elected councils are key players as they steer energy planning and provision and represent the key implicated actor group, the residents buying the district heating. These everyday people typically struggle to enter discussions and decision-making on how the highly infrastructurally invisible (Star and Ruhleder 1996) district heating is organized (Fig. 7.4).

This then results in final sets of considerations about the ecology of actors, taking us back to the "issuefication" and "publics" involved. Having an idea of the actors and structuring conditions helps to estimate whether some actors have built or may build coalitions related to the issue or have interests to politicize the issue (s) involved, and if so, how and to what end(s). Just as importantly, one might want to estimate the reactions from other actors—do they also seek to build coalitions, or do they, for example, seek to downplay the issue's publicness in the public and keep it as a "depoliticized" professional or expert issue? The same goes for the capacity for regulatory action(s) on the issue: what actors are seeking to affect regulations, to what direction,

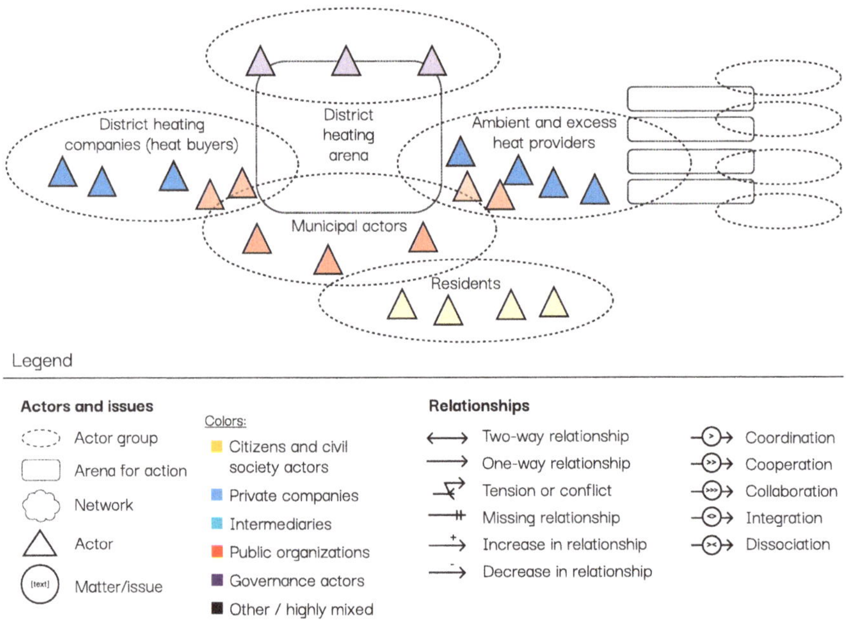

Fig. 7.4 Ecology of actors and arenas in excess and ambient heat reserves in district heating

and with what remits, coalitions, and conditions? (Kivimaa et al. 2018). A further pronounced character of design participation is the relative generality of matters to many or few actors. As we saw in Chapter 5, this can give rise to the need to build "participatory mixes," as any one format cannot address the wide diversity among the implicated actor groups. This can also result in an explosion of meaningful differences. For instance, in design participation for product development, one can quickly run into a situation where a consumer product ought to suit tens of millions of people, and catering for their needs and preferences in a differentiated manner becomes a practical impossibility, typically giving way to seeking to design and define, for example, usages of the product so that they suit decently well the most important user groups without trying to differentiate the offering to the needs and preferences of tens of thousands of diverse actor groups. Figure 7.5 depicts these considerations in relation to actor characteristics in ambient heat use in district heating in Finland.

Design Participation Action for Ambient Heat Resources for District Heating

Informed by these considerations, the chosen line of design participation action can now begin to be elaborated starting from considerations as to what design participation approach (see Chapter 6), or combination thereof, might be suited. In the district heating case, Auvinen et al. (2023) and Auvinen (2024) pursued first a literature review identifying a long list of potential barriers to excess and ambient heat usage and then sent a survey to known heat providers and district heating companies about these barriers. After this, we held a workshop with 38 representatives from district heating companies and potential providers of excess and ambient heat. The barrier list formed based on the literature review and survey was organized into statements in the online platform tool Padlet, and the participants from both district heating companies and from the potential heat provider side were given the task to rate which barriers and enablers they regarded as important for them. This allowed for differentiating which barriers were raised as important by each actor group and what discrepancies remained. The second workshop then focused on the acceptability of policy strategies taken to further the excess heat uptake in district heating. This analysis was used to, on the one hand, argue for what policy measures should be taken in Finland, and on the other hand, it provided the justification for embarking upon developing an industry roadmap or transition arena via serial codesign events. Thus, in terms of the Chapter 6 designer–user configurations, this design participation context was first tackled through quite classic collaborative design measures, which were envisioned to feed into further mid-range transition arena processes, as described in Chapter 4 (Auvinen et al. 2023; Auvinen 2024).

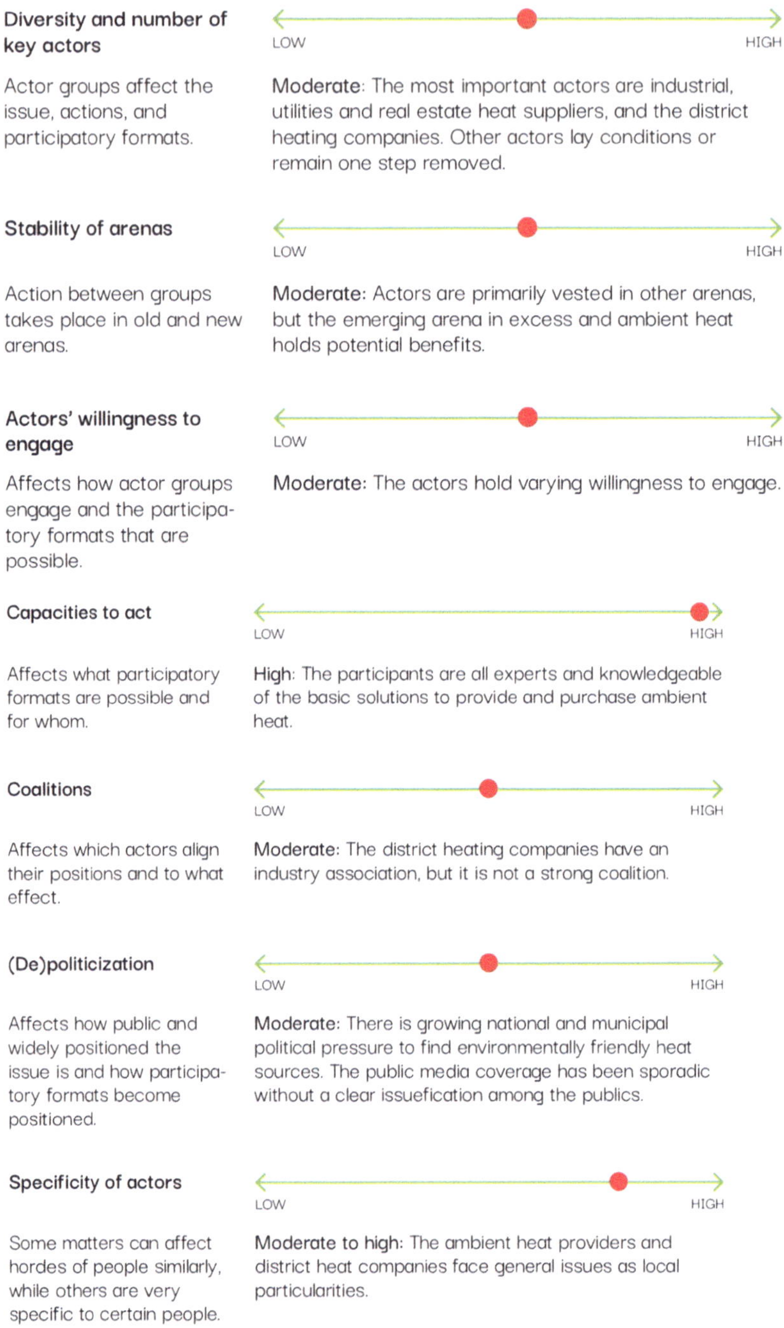

Fig. 7.5 Assessing the characteristics of affected actors, the ecologies they form, and intensities of these characteristics in district heating

Auvinen (2024, pp. 112–117) further envisions that to instill the transition in district heating into accelerated motion, a more throughgoing design participation action program would be needed: an energy transition management program. The initial step in the plan would be running a mid-range transition arena (see Chapter 4) that would include key actors in district heating and ambient heat reserves along with the key regulatory actors, politicians, and powerful actors in the energy sector. Solidifying ambitious visions and committing to a change pathway would next lead to a series of actions and experiments to drive the change forward, coordinated by a national multiparty transition steering group with regular meetings and presence of all the key implicated industry associations and public authorities who hold power and authority in the decarboning of district heating in Finland. The schemata would thus expand colocated collaborative design measures with system change oriented work (see Chapter 5) and design-in-use strategies (see Chapters 3 and 6). The process would include conflict mediation processes and support measures for disadvantaged groups and disproportionately suffering actors and include substantial investments, monitoring, and evaluation, as well as an incentive and compensation scheme.

Design participation can thus seek to solve and set into motion change among actors with relatively low-resource one-time actions such as the two workshops described above. At the same time, it holds potential to support or grow into a wider program for social action, even as the actual realization of such change initiative is in this case, as yet, only a hope. Let us next broaden the above considerations through cases that introduce other common uses for design participation.

Supporting Emerging Phenomena Through Design Participation: Household Hybrid Renewables

The second of the common design participation situations we discuss is one in which a new emerging phenomenon becomes recognized as important and meriting supportive action yet not falling into any existing actor's remit and competency. Examples of such situations can be found in many domains, for instance in how the youth work could relate to youth escaping to digital realms (e.g., Johnson 2013); new outpatient groups becoming recognized by mental health services and NGOs (Kinnunen et al., 2012) and how new customer groups fail to act effectively in the market due to the insufficiency of information provided by suppliers and a lack of needed intermediaries (Murto et al. 2018a, b; Hyysalo et al. 2022). A common measure in such situations is to seek out interested and implied parties to explore and elaborate on the situation and to find doable avenues for supportive action.

Our example case here is **hybrid heating arrangements** in detached houses in Nordic countries. As energy transitions have advanced in households, citizens have pursued additive adoptions of heat pumps, wood stoves, solar photovoltaics, solar heat collectors, and home automation systems (and so on) to opportunistically profit from their different daily, seasonal, price, and labor merits in heating. This has proceeded to a point in which as many as half of the Finnish detached homes may currently have 3–11 heating systems (Numminen et al., 2024; Silvikko de Villafranca et al. 2025).

With respect to **issuefication**, hybrid household renewables have remained under the radar of public and expert attention, likely because the official statistics, public energy counseling, and next to all company offerings assume houses hold just one primary heating source and one supplementary way of heating (Numminen et al. 2023). Yet the hybridization of heating is concretely felt in the everyday life of hundreds of thousands of citizens, most of whom remain unaware that others have also taken on additive energy renovations. The commercial and public actors involved are not offering services or targeted solutions for "hybrids" but instead endorse only single solutions, an integration of which thus remains with the adopters who then seek help from their peers locally and over the internet (Hyysalo et al. 2022; Hernberg and Hyysalo 2024; Silvikko de Villafranca et al. 2025). We shall next explore the dimensions of hybrid heating developments to gain a better grasp of this context as a site for design participation of seeking to shape the development toward public good.

To again gauge the "ballpark situation of action" for hybrid renewables in houses, let us tentatively arrange the contextual elements into a situational matrix without yet imposing assumptions about their mutual relations or relative importance (Fig. 7.6).

To then again start clarifying the presence and gravity of **structuring conditions** in the case, let us turn to "sliderizing" the same conditions as we did with the district heating case (Fig. 7.7).

The preliminary assessment of involved structuring elements indicates that the topography in the situational matrix remains mostly "flat" without many or high structuring effects in place (Fig. 7.7). This suggests that iterative ways forward are likely possible in furthering hybrid renewables and improving the support that citizens would want with their energy arrangements. With these parameters in place, one can proceed with approximation as to what the actor positions and coalitions are and how fixed they may be, as well as what agendas different actors may have and issues they strive for.

In hybrid renewables, an important actor group is the homeowners that have installed hybrid heating arrangements. Their exact number is not known, but studies and statistics suggest there are 200,000–500,000 of them, and there are important differences between them regarding whether the heating arrangements are planned or have evolved over time; whether they include fossil fuel heating in a remnant role; and how integrated they are and how much technology modification there is (Numminen et al. 2024; Silvikko de Villafranca et al., 2025). The second actor group is suppliers and installers of different RET systems: air-source heat pumps, ground-source heat pumps, solar PV, solar heat, pellet and woodchip burners, and home automation systems. The third actor group is public energy counseling bodies: national,

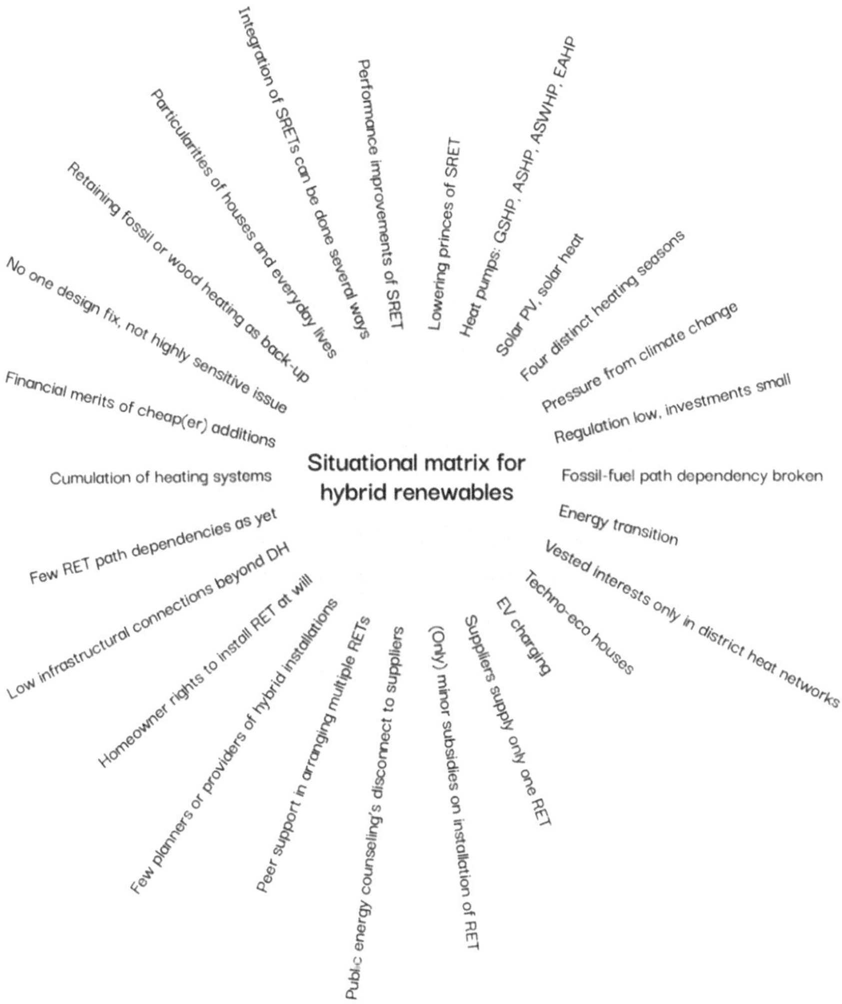

Fig. 7.6 Tentative situational matrix for hybrid renewables

regional, and various municipalities. The fourth actor group is regulatory bodies: building regulation officials and Finnish energy authorities who have a say over installations and protocols in protected buildings, larger installations, and prosuming energy back to the grid, respectively. The fifth and final actor is Statistics Finland and the building registry as a bridge between regulation and political decision-making that rests on (outdated) statistics about the rapidly changing phenomenon. Figure 7.8 depicts this ecology of actors, key arenas, and disjunctions between the actors.

The tentative situational matrix and depiction of ecologies of actors, their social worlds, interactions, influences, and interstices (or "gaps") in place, we

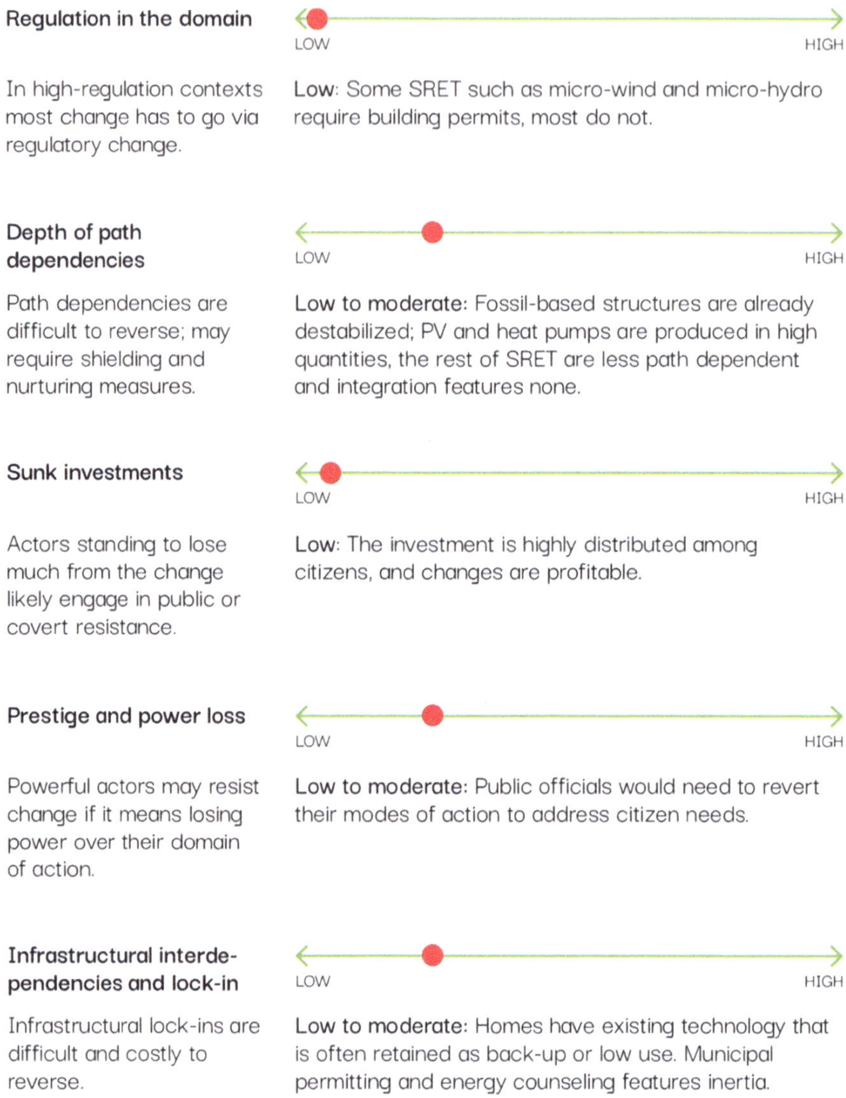

Fig. 7.7 Assessing the presence of structural sociotechnical conditions for hybrid renewables

can start to expand onto a more fully fledged understanding of the situation for design participation. The issue is that we often cannot know it all from the outset, nor do we have time and resources to pursue an all-encompassing social science assessment prior to departing for participatory action—“paralysis by analysis” or “by default overengineered participation” is not what this

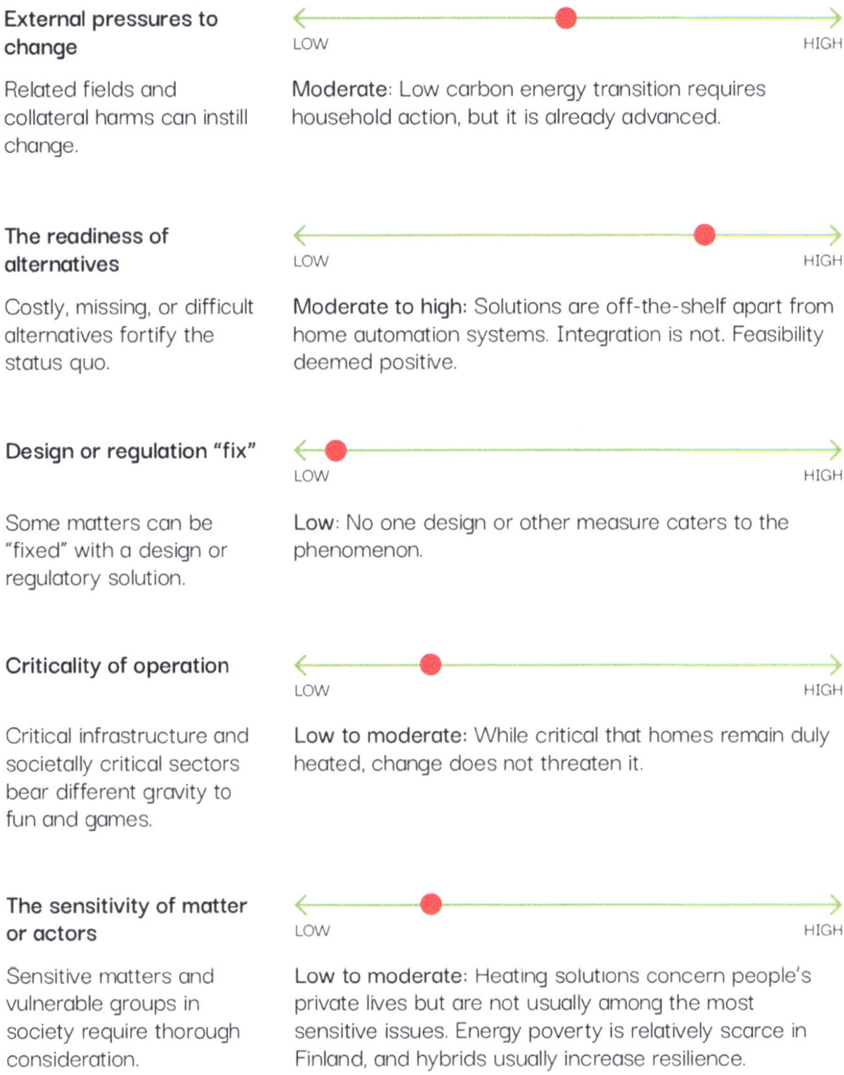

Fig. 7.7 (continued)

chapter's scoping guidance is a recommendation to do. Rather, the implication is that one must take initial bearings into the matters, issues, and actors at hand and then devise the design approach so that it matches the uncertainties involved. In some cases, this can be a clear-cut one-time workshop design, and in others a more prolonged participatory engagement, and building infrastructure for participation is the requisite for making a positive difference and to better reveal the conditions for change. Figure 7.9 condenses the same

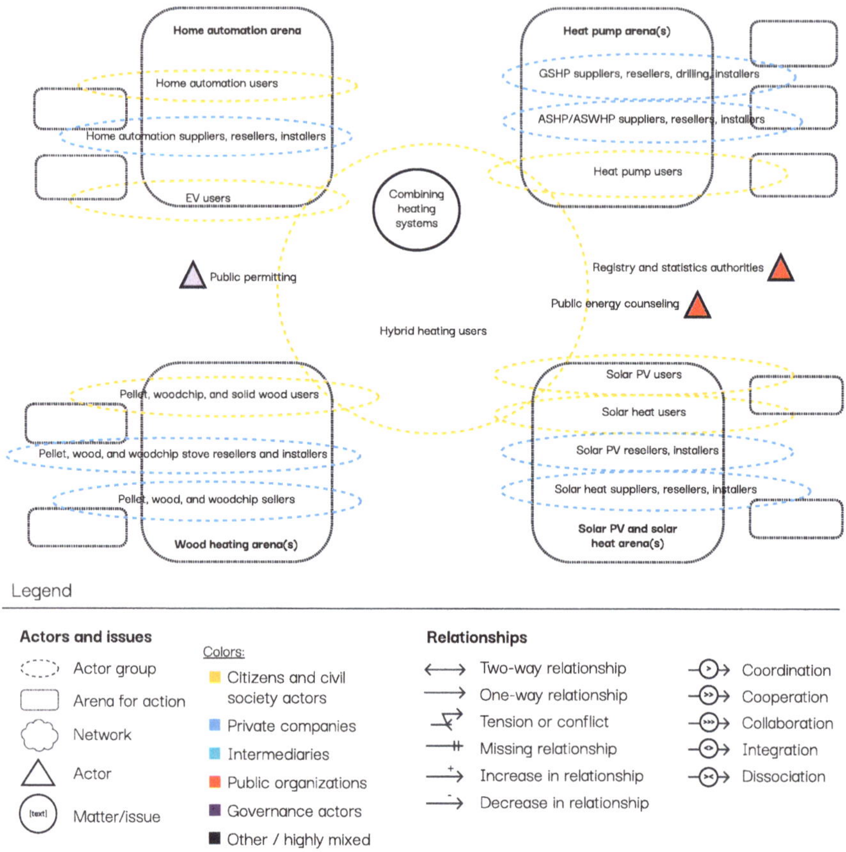

Fig. 7.8 Ecology of actors and arenas in hybrid heating. The hybrid heating arena is only weakly structured, consisting primarily of users of RETs that have their own arenas of development, retail, and use. Permitting and energy counseling and official statistics all acknowledge the phenomena exist but do not cater for it

considerations about actors and their interrelations that we had for the district heating case.

What kinds of design participation do these assessments suggest for hybrids? Across the chapters of this book, we have underscored the intertwined considerations related to relevant participants and participations; assessment of the best suited design participation approach(es); the attainment of human, financial, and organizational resources; the participation formats and arrangements and more specific material and social mediation within those; working out the procedures and temporalities involved; clarifying the sought and likely outputs and output work for recording the results and their refinement; planning design iterations and expansion strategies to the design participation; and ultimately, iterative translations between these to reach well-working design

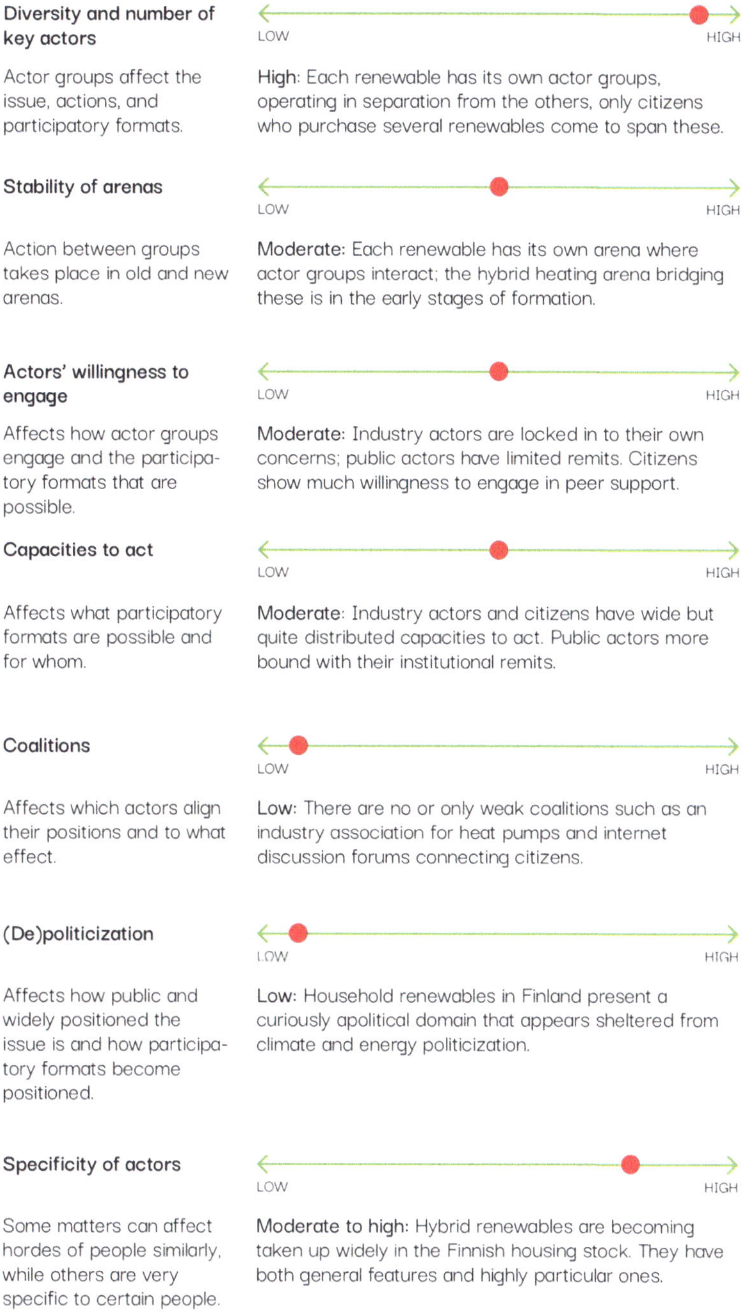

Fig. 7.9 Assessing the characteristics of affected actors, the ecologies they form, and intensities of these characteristics in emerging hybrid heating

participation arrangements. The details to which all these are sensible to work out from the onset varies greatly. In invited participation formats, the hosts can do extensive intermediate designing (cf. Chapter 2 workshops), whereas the more distributed and emergent forms of design participation that rely on participants' own initiative and organizing are less amenable for similarly detailed preplanning.

Against this backdrop, design participation action for hybrid renewables would likely have its best focus in targeting the disconnects and unawareness of the phenomena among the actors who could better support it. Users exchange actively about hybrid solutions on the internet forums in some disconnect from energy counseling and again in disconnect to technology suppliers and installers. Nobody holds accurate information of quantities of hybrid heaters, let alone what they, and the people likely to next buy hybrid heaters, have as pain points, what they would need and value in terms of public support, and development of business offerings.

A potential starting action would be to invite members from each of these actor groups into a day-long workshop to discuss the challenges and possible solutions and development needs related to increasing hybridization of heating arrangements in households. The workshop could be structured to three parts: challenges, ideas, and solutions and development actions forward. A good way to proceed would be to run small groups on different sub-themes in each part, followed by sharing and general discussion. To make the workshop effective, physical or digital canvases to collect the statements are likely needed as well as time to prioritize the most important matters and ideas. The multi-actor workshop could be rendered even more effective via discussions or interviews with each actor group beforehand to identify their key issues and concerns, or, alternatively, smaller within-actor-group workshops could be held as a preamble.

While such starting action can be planned in fair detail, its role would foremost be to start more enduring and exploratory participatory actions that could really make a difference. One avenue to be explored among the participating actors is whether and how public energy counseling and private energy consultants could participate in the online communities where hybrid heaters de facto source the help they need (Silvikko de Villafranca et al. 2025). Another more open-ended avenue would be to trial experiments on if peer community formation could be moved to a higher gear by setting up a dedicated hybrid heating section(s) in the most wide-ranging of the internet forums and/or other social media. Such peer community build-up is however highly contingent and dependent on whether interested participants find the community and whether people take on volunteer moderator roles and it would thus need to follow more responsive design-in-use tactics. Yet another complementary and more commercial route for action would be to seek, for example, a series of meetings with commercial actors to explore if they could better support hybrid heating households through partnerships, while recognizing that their business model typically rests on just one technology.

Shaping the Direction of an Emerging Industrial Field: Desirable Hydrogen Pathways

The final case examines the use of design participation in shaping the direction of industrial fields to promote sustainability transitions. As discussed in Chapter 4, such codesign for transitions presents a new and growing avenue for design participation often promoted under the banners of transition design (e.g., Irwin et al. 2015; Ceschin and Gaziulusoy 2019), often in the face of complex and contested industry structure renewal.

The case we examine in this regard is green hydrogen production in Finland, which has been spurred into motion by the advancing energy transition induced by climate change. As northern European energy transitions have moved from their early phases to society-wide reconfiguration of energy production and use, the future visions have increasingly come to emphasize the conversion of renewable intermittent electricity to hydrogen through electrolysis (Finnish Hydrogen Cluster 2023). Initially, such green hydrogen can be used to replace the fossil-fuel-based hydrogen used by various industries. Once the capacity grows, the hydrogen allows "power-to-X" for synthesizing methanol, ammonia, and other compounds to replace fossil fuel usage in hard-to-electrify sectors. As importantly, the electricity used in hydrogen production can be a means for balancing the peak supply in the electricity networks resulting from increasing shares of intermittent wind and solar production, thus helping to retain power-grid stability. Green hydrogen production is consequently envisioned to be conductive for profitable growth in intermittent renewable energy installations, which in sparsely populated countries such as Finland could result in energy exports and energy-induced reshoring of industry. In these hydrogen economy scenarios, the massive deployment of wind, solar and hydrogen production leads to investments in power production and power-hungry industrial investments on the scale of tens of billions of euros of annual turnover in Finland alone, either as exports via European pipelines or through build up of local heavy industry "valleys" near wind and hydrogen availability (Gasgrid Finland 2024).

In terms of design participation from the perspective of actor(s) aiming to further energy transitions and their democratic governance, the field is currently taking shape and could result in many societally consequential development paths. Hydrogen production facilities, pilot plants, pipeline siting, policies, and subsidies are becoming envisioned among the experts and key energy-related organizations. Governmental energy roadmaps are being commissioned, and an ecology of actors vested to some degree in the hydrogen developments has taken shape. However, hydrogen developments have remained (at least at the time of writing) much slower to materialize than envisioned. Perhaps because of this, they have not become issuefied into political or public media discourse in Finland. Envisioning, lobbying, and debate is happening among the experts, investors, and regulatory bodies, yet there is an awareness that the hydrogen pathways are becoming increasingly a social and public and "issues" over the next few years, with the potential to become politicized nationally and in various affected localities. We shall below explore the dimensions of the hydrogen developments to gain a better grasp of this context as a site for design participation from the perspective of seeking to shape this development toward public good.

For the hydrogen developments, an early situational matrix could thus look like in Fig. 7.10, depicting at least the conditions and actor groups that are saliently present but not yet assigning interrelations or relative importance among them (Fig. 7.10).

The preliminary assessment of involved structuring elements (Fig. 7.11) makes it evident that in hydrogen developments there are powerful actors, big investments, critical infrastructures, and established technological pathways present. This saliently high sociotechnical topography thus presents quite a different context to the relatively flat structuring present in, for instance, hybrid renewables.

Let us next we move to assessing actors in more detail to complement the situational matrices and to outline more dynamic mapping of the ecologies of actors involved. In hydrogen developments, the connections and interlinkages between the actors are still emerging but also quite complex. The key actors are industry companies, most listed within the network of "Finnish hydrogen cluster" with varying interests in electrolysis plants, transmission grids and facilities, or storage and chemical processing of hydrogen. Another set of key industry actors are big energy users, particularly those using hydrogen directly in their processes as they constitute the first phase demand for hydrogen. The third industrial actor group is the wind power companies and prospectors who will need hydrogen developments for absorbing the increases in wind power production both onshore and offshore. The fourth actor group is the grid operators, Gas Grid Finland and Finngrid for electricity; the fifth actor group is the regulatory actors, the energy authority and the regional business and environment authorities (ELY), who do industrial permitting processes. The sixth group of actors is the other policy actors and politicians who are implicated in the industrial development that hydrogen development entails. The seventh actor group is the research institutes and universities providing research and development inputs and workforce for hydrogen development, under the banner of the Finnish hydrogen innovation network. The eighth group, and not yet involved actors, is citizens implicated by the hydrogen developments: those living next to and in the municipalities of additional electrolysis plants and wind parks and potential new hydrogen pipelines; prosumer citizens wishing to utilize their homes and allotments for demand response systems and for PV production; and consumers to whom the added wind power made possible by hydrogen could provide lower and more stable energy prices unless the industrial facility additions or export pipelines take away the domestic advantages. The potential and actual arenas between these actors are many. Figure 7.12 presents the tentative mapping of this ecology of actors,

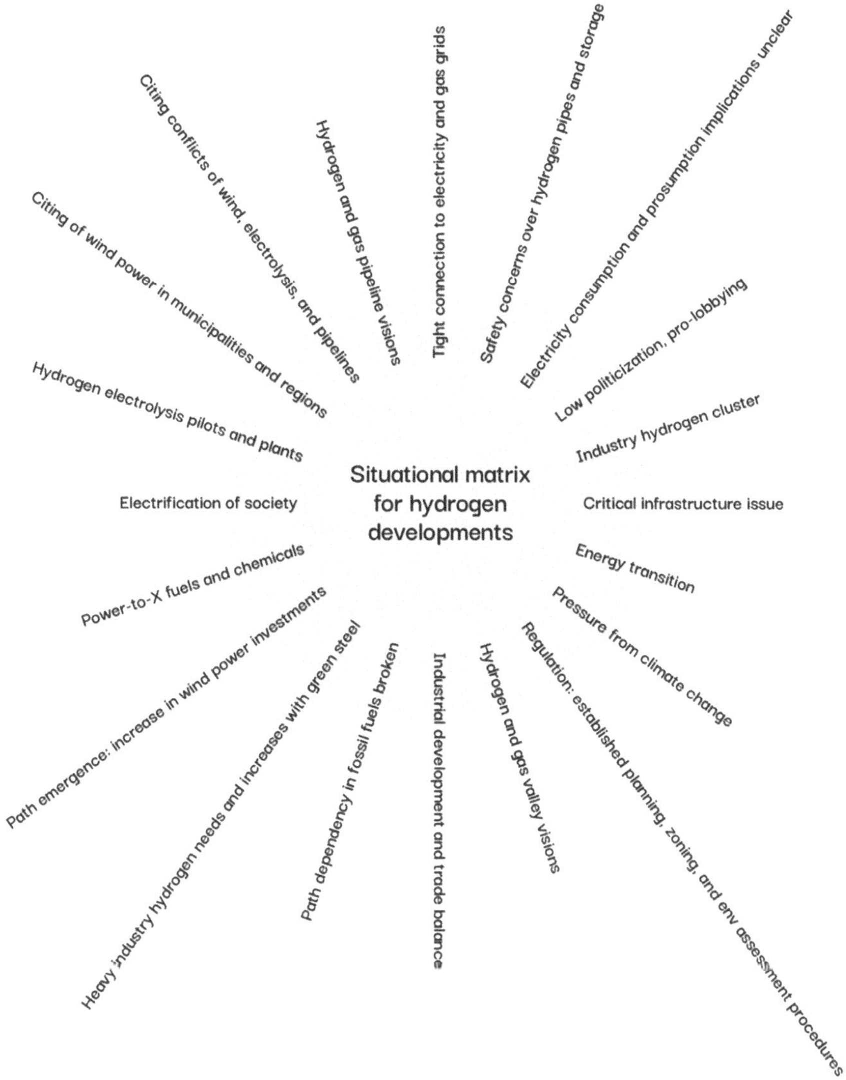

Fig. 7.10 Tentative situational matrix for hydrogen developments

and Fig. 7.13 presents the characteristics of affected actors, the ecologies they form, and intensities of these characteristics in hydrogen pathways.

The tentative situational matrices and depiction of ecologies of actors, their social worlds, interactions, influences, and interstices (or “gaps”) in place, we can start to expand onto a more fully fledged understanding of the situation for design participation.

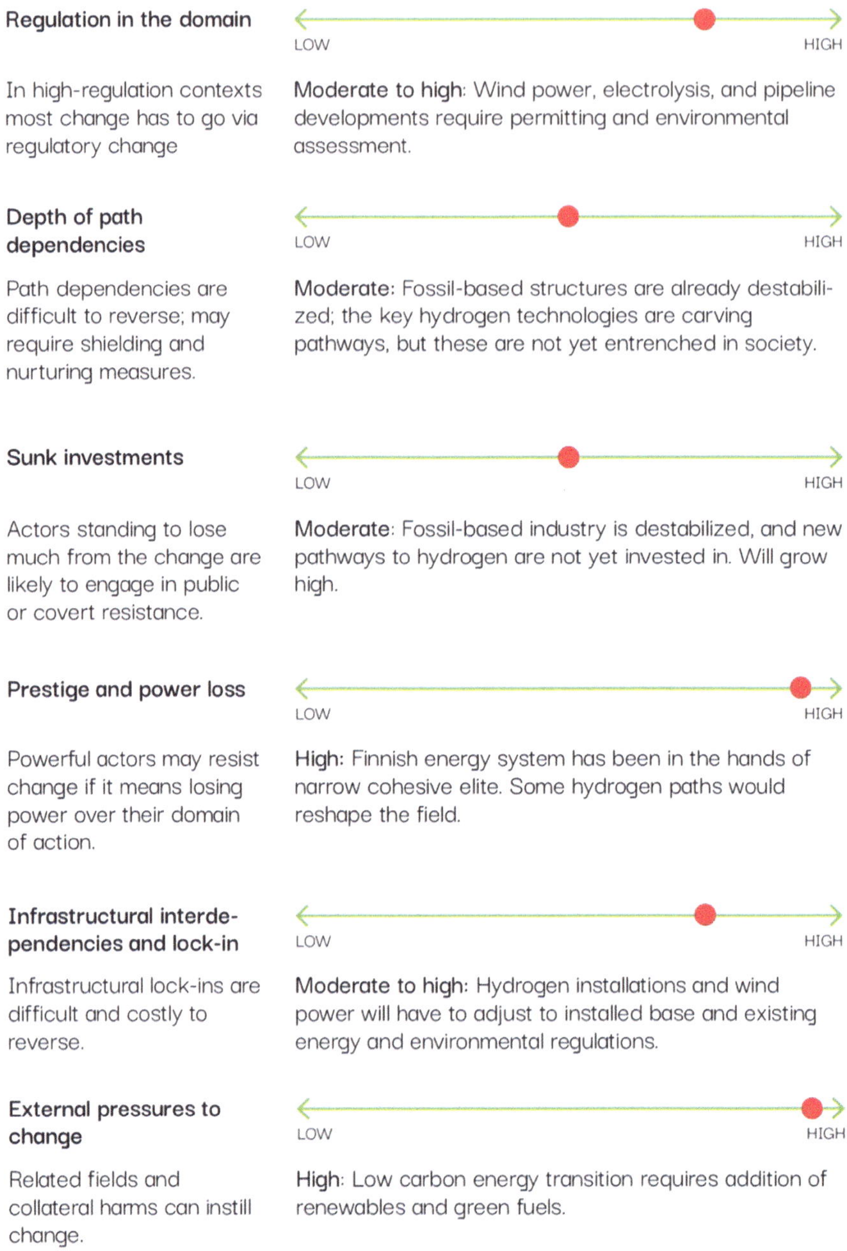

Fig. 7.11 Assessing the presence of structural sociotechnical conditions for hydrogen developments

The readiness of alternatives

Costly, missing, or difficult alternatives fortify the status quo.

Moderate to high: Key hydrogen technologies are at pilot level but improvement and scale-up ongoing; feasibility features uncertainties.

Design or regulation "fix"

Some matters can be "fixed" with a design or regulatory solution.

Low: No one design or other measure caters to the phenomenon.

Criticality of operation

Critical infrastructure and societally critical sectors bear different gravity to fun and games.

High: Concerns that critical infrastructure, energy transition, and industrial investments hinge on RET addition.

The sensitivity of matter or actors

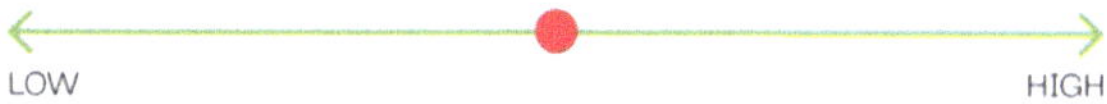

Sensitive matters and vulnerable groups in society require thorough consideration.

Moderate: Citing industrial installations has effects on local people; RET addition affects energy prices and investments by citizens.

Fig. 7.11 (continued)

For steering hydrogen developments in Finland, the backdrop is one of strong industry interests and economically and societally powerful actor groups. These make it likely that carefully organized and quite formal participation processes would be required to shape or even inform the developments. The design participation needs legitimacy in the eyes of key actor groups in addressing core concerns and uncertainties involved, and while, for instance, in the hybrid renewables case one likely could proceed iteratively and somewhat informally, in the participation processes in hydrogen developments, one would need more formal background work and higher elaboration of methods and participation setups from the onset. It is thus highly likely that small-scale or informal efforts toward widened participation may not gain interest, legitimacy, nor societal effects among the key actor groups. The situation is in several respects similar to the transition codesign we described in Chapter 4, where a mid-range transition arena was pursued to foster learning among the participants about the possibilities and implications of different sociotechnical pathways. One possible point of departure is thus to refine this approach

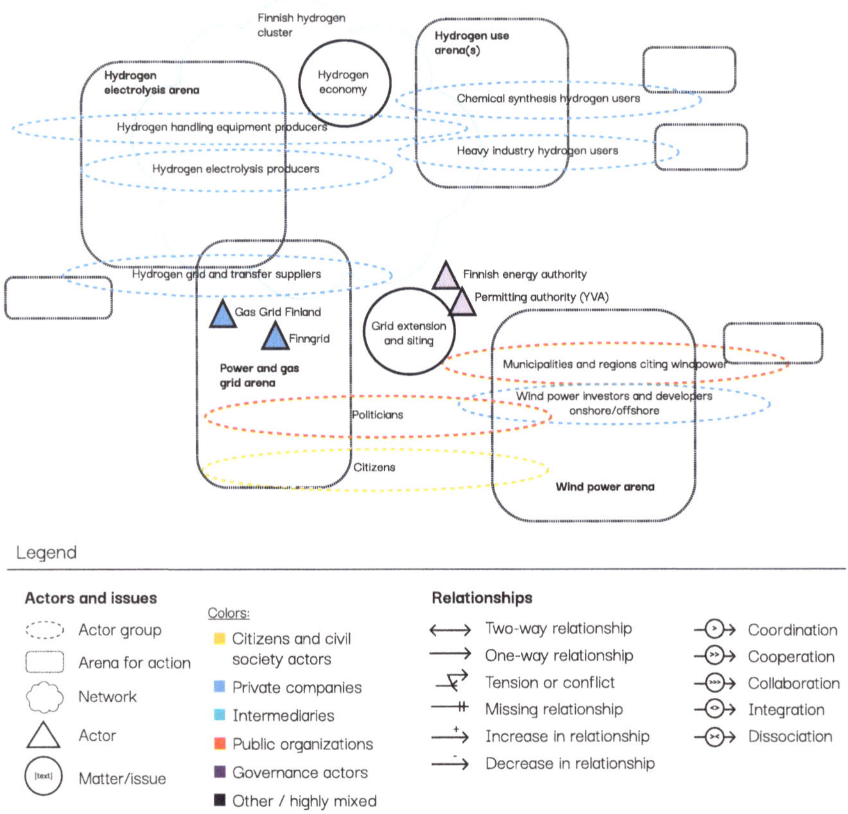

Fig. 7.12 The tentative ecology of actors for Finnish hydrogen development. The ecology of actors operates in loosely coordinated electrolysis and hydrogen use and wind power arenas

further. In terms of organizing, it could be sensible to co-organize it with a legitimate actor such as the Finnish hydrogen cluster, hydrogen innovation center, or Finnish GasGrid. Regarding participants, it is likely that a diverse mix of 15–30 participants from the key actor groups would be the best, in keeping with heuristics used in MTPT arenas before (see Chapter 4). The concrete workshops, their pacing in the calendar, and timings within the workshops and contents would need to be iterated with the co-organizer. Also, the arrangements to record the workshop interactions and outputs can in this schema be adopted or somewhat adapted from previous MTPT arenas. Yet the participatory envisioning of hydrogen development pathways appears to also call for additional new means and tools for work. The existing MTPT tools described in Chapter 4 do not open up considerations of relevant actors and interrelations, and importantly for the foreseeable hydrogen developments in Finland, the changes in the ecology of actors would need to accompany the

Fig. 7.13 Assessing the characteristics of affected actors, the ecologies they form, and intensities of these characteristics in early hydrogen developments

growth of the area—such analysis could then be added in MTPT as an additional tool, and/or a new phase in the mid-range pathway arenas. Indeed, the actor groups and arenas mapping used in this chapter are something that we envision could provide this needed new means.

Given the expert-driven processes to date and highly implied nature of citizens, it would appear sensible to somehow have a focus point toward considering the implications for citizens in different pathways and their phases. This could potentially be included in the ecology of actors mapping, and the design challenge would remain how to do that in practice.

Chapter Conclusions

In this chapter, we have outlined one way to approximate the ballpark situation of action and the ecology of actors involved in different design participation contexts. From here, each case would proceed into detailing the design participation actions to be taken and, as indicated in Chapters 2 and 5, this would typically result in considerably more refined considerations and estimations of the contexts involved and of how the further actions in design participation could actually be best fostered.

In academic projects and highly important projects, the context assessment would become considerably more nuanced and detailed involving talks and interviews with trusted actors knowledgeable of the participation contexts, whereas in small practice projects, the assessment might remain largely heuristic and rely largely on, for example, client-provided information.

The other item that the book as a whole and the three energy transition cases in this chapter underscore is the considerable variety in design participation contexts and the likely suited courses of action—no one method, method mix, or approach likely fits all contexts well. An attempt to proceed with one favorite method likely introduces a version of the "whole world looks like nails if you have (just) the hammer in your head" parable. In turn, this easily results in overly expensive, underwhelming, or just ill-fitting participatory measures. The resulting shortcomings, in turn, easily undermine the subsequent use of participative approaches overall. Our main message thus becomes that one not just should, but one must, take one's bearings and devise one's approach to design participation carefully enough to at least know how carefully they need to be taken. And in so doing, one should consider a wider range of available courses of action than just one default approach, be it participatory design workshops, transition arenas, developmental workshops, deliberative minipublics or other (however impressive) ways of working. It is vital to approximate the context sufficiently to be able to gauge *how* the design participation action can be meaningfully pursued. For instance, while the strong sociotechnical structuring and economic and power interests salient in the hydrogen development quickly point to the need for a legitimate formal participatory approach such as mid-range transition arena, they also imply more. The mid-range arena is likely to require a fair amount of pre-work to succeed as well as considerable resourcing and facilitator skill—running it with just a small or ad hoc designer team that might work in many other transition design projects would here run a strong risk of failure (Irwin 2015; Ceschin and Gaziulusoy 2019).

The final implication from the considerations in this chapter is that design participation can be used to catalyze, champion and steer social action, but wide social movement mobilization aside, it cannot be expected to alone solve thorny social issues in contexts such as energy transitions. It can thus be thought of as a specific intermediary action that can steer and catalyze other actors' positions and capacities toward desired directions (Kivimaa et al. 2018; Hyysalo et al. 2022; Hernberg and Hyysalo 2024), often importantly clarifying what those desired directions might be.

CHAPTER 8

Conclusions: What Design Participation Can Offer and How to Overcome Its Pitfalls

We opened the book by arguing that design participation presents tried and doable inroads toward environmental and social change. This is because many environmental and social issues are structured so that multiparty knowledge coproduction, elaboration of design problems and solutions, direction search, and more widely informed decision-making are requisites for moving toward sound action. Participation matters.

The matters of participation, in turn, cannot be adequately handled in a "just add users and stir" manner, as Muller and Druin (2009) nicely rephrase the old feminist dictum (Harding 1995). Let us be clear about this. Involving the affected people in design and stirring does on occasion work at least to some length (for a documented example, see Lehenkari and Hyysalo 2002, 2003), but most of the time the matters of participation require more considered ways of acting. And while the matters of participation are elaborated across several bodies of academic literature, the predominantly prescriptive and dispersed nature by which this is done does not straightforwardly carry through to informing design participation—as it is practiced or even academically. We would thus like to close the book by explicitly elaborating how the book stands with respect to some commonly endorsed academics' and practitioners' positions toward design participation.

Our insistence on design participation work and complementarities between different approaches to design participation are underpinned by taking design participation as one area in which *democracy is rehearsed* (Pateman, 1970). Design, planning, innovation, and technology development have traditionally been expert-driven areas of society, in which citizen participation has remained rather limited even in democratic societies. As this has changed particularly during the last two decades—making participation part and parcel of how our current design, innovation, and transformation activities operate (Hyysalo,

S. Hyysalo, *Design Participation*,
https://doi.org/10.1007/978-3-031-95437-5_8

Jensen et al. 2016; Chilvers and Kearnes 2016)—it is important to keep improving these practices both for the organizers as well as for the citizenry that participates. In most sectors, there is a routine backdrop as to how participation, design, and planning have been intertwined, such as in environmental governance where the previously developed participation formats in legislated hearing procedures and appeal processes have recently become paired with forms of knowledge coproduction (Turnhout 2020; Miller & Wyborn, 2020), opening up space for experimenting with and rehearsing design participation in, for example, planning and sectoral transformation, as we saw in Chapter 4. As democratic governance overall is under increasing attacks and threats, innovating in and rehearsing participation are increasingly important to foster wider and more deeply rooted democratic engagements.

Across the book, we have been clear that *there is no one sufficient nor supreme method to design participation that can be readily deployed across different settings*. The elaboration of methods and seeking to justify some method's supremacy over the others continues as perhaps the most widespread genre in design participation-related literature. Methods and approaches tend to have their ups and downs in popularity in different application areas. At the time of writing this book, deliberative democracy mini-publics have mainstreamed into high popularity, owing to years of academic elaboration, instrumentalization for conducting them in practice, and presenting novelty value in extending democratic engagements (Voß and Amelung 2016). And while deliberative mini-publics clearly do have their merits, times, and occasions, overextending their application is not the best of ideas. For instance, the random sampling-based participant selection for mini-publics is very unlikely to ever result in a participant group that could develop high-quality insight on such emerging issues that remain ill-elaborated to organizers and experts alike. The capacity to hold high-quality elaboration on such topics is concentrated to a narrow band of citizens with a high topical knowledge base, and they cannot be identified and enrolled through random sampling.[1] The same goes for participatory efforts that wish to give direct design control over to participants, as random sampling is also highly unlikely to result in a very design-able mini-public (see Chapters 1, 2 and 6). Furthermore, if deliberative mini-publics are taken as the only legitimate way to pursue democratic participation additional to institutions of representative democracy, their requisite

[1] Notice that this does not mean subscribing to a common objection by design practitioners that not-design-trained people are not competent at design and innovation. While most people are not competent at design and innovation in most situations, the requisite competencies can almost always be found among some citizens. Whether it is worth embarking upon the often-arduous search process and careful relevance framing of the issue to attract them is a different matter and boils down to the selection principles and how the selection work is carried out.

resource intensity will quickly result in forsaking public participation as unviable in most cases.[2] Conversely, deploying ways to involve user innovators and lead users in participation holds great promise for innovation and generations of future insight, but if it is the only format used it is likely to remain quite narrow in widening democratic opportunities (Hyysalo 2009). And again, also the costs associated with identifying lead users would close off participation in most cases. In all, regardless of how many great merits a method has, or how personally fond one might be of any one method or approach, any one of them will necessarily cover well only a subset area of design participation, as we elaborated in Chapters 5 and 6.

Similar to methods, only *elaborating principles will not be enough.* The elaboration of principles and prescriptive political science positions to inform participation has been a central pursuit not just in political science but in design and planning as well (e.g., Bjerkness 1987; Ehn 2008; Escobar 2018). Across the chapters, we have underscored how realized participation is always an emergent outcome from the participating subjects, the objects of participation, and the formats by which it is pursued (Chilvers and Kearnes 2020). Furthermore, in real-world settings, participation is necessarily permeated by mundane and strategic work of different types that affect whatever is made of, and whatever can be made of, participation and what outcomes it can produce (Hyysalo, Jensen et al. 2016, 2019). Participation is a considerably more ambiguous and practical-skill-requiring endeavor than most academics would care to admit. Democratic practices matter as much as the democratic principles.[3]

We have also insisted that different disciplines have elaborated on important premises regarding high-quality participation. But these elaborations do not cohere to a single unified view, and the tendency to assert that "the *deeper the participation the better" sounds far simpler than it is.* For instance, the legacy of ideas elaborated in various participation ladders, typically owing to Arnstein's (1967) ladder of citizen participation, lives strong in tens of different ladders,

[2] Notice that this does not mean subscribing to a common practitioner objection that participation is too resource-consuming per se. Surely, participation for participation's sake is a waste of resources. And design participation will require resources—sometimes considerable resources—and this expenditure better be weighted against likely attainable gains. These gains are often manifold, ranging from higher ownership, legitimacy, design contributions, and marketing gains to improving designs and decisions. Some have direct monetary benefits, and others have indirect ones (Johnson et al., 2014; Pollock & Hyysalo, 2014).

[3] Notice that this does not mean subscribing to the common for-or-against participation stances among design practitioners that reduce participation fundamentally to an ideological commitment. Participation is not a sole or sufficient value for transformative endeavors, nor is it necessary in any and all transformative pursuits. In practice, design participation is an anxious mix of substantive and instrumental values dependent on its mundane realization. A wording we used in the introduction was that design participation needs to be pursued "for real": for real problems, skillfully, and in earnest to become effective. This is in contrast to deploying it as a value of its own, as an additional nicety, or as a standard tick-the-box procedure.

including for instance the International Association of Participation's spectrum of citizen participation (IAP 2023; cf. Kelty 2020). No doubt, it is important to differentiate between manipulative and therapeutic deployments of participative techniques from participation premised on informing the organizers and again from those where participants have increasing decision power (Arnstein 1967). But the decision power axis is just one of the "depth dimensions," or discursive justifications, involved (Irwin et al. 2013). Irwin et al. (2013, pp. 126–127) draw attention to several commonly employed registers to discuss public participation, in effect always problematizing it as more or less representative, as more or less impactful, as more or less conductive to participants being able to articulate their opinions, and as more or less supportive of other democratic endeavors. On top of this, in Chapter 2, we underscored the importance of how deeply elaborated the issues and matters are for the participants—superficial understanding will result in superficial decisions—and how the requisite depth may take years to gain. Participants' ability to directly design and alter the outputs of participation—their design control—is similarly another key depth dimension as it affects how directly the participants can shape design or planning (in contrast to filing requests to others who then make the eventual design decisions and who can always use design control as a reason to endorse, re-interpret, or reject any given request). In Chapter 3, we further underscored how the "run-time" of participation affected how deeply participants could affect the co-evolution of designs and practices, as the consequences and possibilities of design decisions become observable and act-able only during use time. In Chapter 4, we went on to underscore how the depth of design participation over sociotechnical change owed to the capacity to collectively span different knowledges and positions. The extent of knowledge integration and mitigation of power asymmetries (and so on) are equally justified vantage points for determining the depth of participation. The net effect is that the "depth" of participation is not a dimension. It is a result of multidimensional and complex engagements in which gains in one dimension often mean a loss in another. The overarching position informing the book is thus viewing the depth of participation as a profile rather than even a sum variable and seeking to give means to assess and make reasoned (tradeoff) decisions about it.

This gets us back to the importance of *skilled work in and of design participation*. Precisely because design participation is necessarily a context-dependent and tensioned pursuit between multiple competing "goods," its skilled achievement is not reducible to one-time design or participation engagement but depends on rehearsing the stock of knowledge and skill in designing for, setting up, running, reflecting on, and building on design participations. This furnishes those carrying out the work in anticipation of how other actors involved and implicated by the work are likely to respond and how those responses are patterned, allowing for skilled and responsible decisions and tradeoffs between competing goods to participation (cf. Irwin et al. 2013). This is particularly so, as work in design participation involves a wider combination of different kinds of practical actions than just "designing" and "sharing," as people not as yet versed in it may perceive it.

Drawing from studies of work in collaboratively organized settings (e.g., Strauss 1993; Engeström et al. 1996; Verheig et al. 2016), we outlined framing work, alignment work, competency building work, selection work, intermediate design work, collaborative (design) work, and outcome work as typically necessary for the realization of design participation. The intertwinement and anticipation of these work types while pursuing another are necessary for the design participation to carry through. At the same time, the different types of work also hold independent existences. The skills and people involved are seldom in a one-off performance but deployed in other projects and settings as well, with similar or different interconnections, and because of this, they may even be carried out by different people altogether, furthering the work practice across sites. Work further points to the ongoing and unresolved character of practical action: none of this work "solves" design participation for good but elaborates and transforms the problem of design participation in a given project (Marres 2012, pp. 40–58; Kelty 2020). Attending to work necessarily means traversing from the minutiae of work performances to these being a particular enactment of skills, materials, and images developed over time, which are, in turn, occasioned in organizational and institutional settings and intersecting political economies and ecologies and actors and arenas formed between them (Strauss 1993; Star and Strauss 1999; Clarke and Star 2003; Engeström 2000). As a conceptual register, work thus pushes our attention beyond any one layer of how things get done, inviting us to remain observant to both the specifics in its enactment as well as the historically wider constituents that are being enacted in any one occasion of work performance (Akera 2007; Clarke 2005), as we insisted in Chapter 7.

The turn toward democratic practices further entails that *high-quality requirements for participation can, in fact, also act as exclusion criteria and hinder the extent of participation.* Whatever we may think about the "depth" of participation being comprised of, its corollary has been the academic zeal to being able to elaborate it, if not produce it in practice, be it in terms of "legitimate," "genuine," "true," "encompassing," "robust," "effective," and "feasible" (and so on) participation. This, however, assumes that the object (issue, matter) of participation is such that the participants want to (or often prescriptively imposed on as should want to) participate in it adequately and deeply. And herein, we tumble to a variant of a well-elaborated political science problem that has been used to justify representative democracy over direct participation since the times of John Locke: people being too busy to perform the duties that full-fledged participation in the political community would require (Marres 2012, p. 67; Pateman, 1970). If we just focus on participation, societies are teeming with well-justified pursuits to which we should participate as citizens, and we can only be quite selective in what we participate in and how. Most participation in society, and certainly most design participation, is not the prime "going concern" for most people who legitimately could, or should, participate in it.

Yet, while we should duly acknowledge the merits of representative democracy, the position of this book has been that it is not helpful to treat this as an "either-or" question between representative and direct forms of democracy, as prescriptive political science would have the propensity to do. The complementarities of representation and direct participation are evident particularly regarding design, planning, innovation and sociotechnical transformations that continue to pose marked difficulties for representative democracy. In Chapter 5, we noted how deep design participation could be "mixed" with quicker and fleeting digital and event-based participation formats for the same public matter. These participation mixes allow for different versions of the design issues and formats to coexist and make participation practically relevant for wider segments of the public, beyond the highly interested and the very time-rich (cf. Marres 2012). More generally, whatever participation format is used, it will perform not only inclusion but also exclusion of some implicated actors, and not just in principle but through whatever its concrete operationalization is. The relevance work and selection work can seldom produce full containment of participation regarding the usually large and diverse implicated actor groups.[4] This in mind, whatever a legitimate profile of participation may be, it is a relational one taking place within ecologies of actors and arenas (Hyysalo 2021) and ecologies of participation within them (Chilvers et al. 2022).

Continuing from the last points above, the literatures pertaining to design participation continue to veer disproportionally to the perspective of designers or other organizers of participation. Yet all ***invited and coproduced participation owes to distributed participation***. In the introduction and across chapters, we have tried to open that space into emphasizing how critical the competencies and interests of participating people are to whatever participation is held or whatever codesign and coproduction with users is sought. The bottom-up design engagements by citizens have been aptly characterized as participation that is distributed among the participants, materials, settings, and formats of engagements (Ryghaug and Skølsvold 2021), and it mushrooms among the citizens who pursue design engagements in their work and everyday lives.[5]

[4] Notice that acknowledging the need for participation mixes does mean subscribing to the common practitioner lament that participatory design and decision-making will veer toward the lowest common denominators and eventually satisfy none. Surely, if the relevance work and selection work are premised on just the lowest common denominators, or intermediate design work is geared toward the lowest common denominators, there is a real risk that the processes and outputs of participation, too, will remain unsatisfactory and unperceptive.

[5] Admittedly, it is difficult to write for designers and other organizers and practitioners without emphasizing their perspective, and it is not a coincidence that my previous book (Hyysalo 2021) focused on the citizen perspective at the expense of design and organizational perspective.

Digital peer production has opened dramatic new vistas for design participation during the last three decades (Benkler 2006; Kohtala et al. 2020; Hyysalo, Jensen et al. 2016). The last decade has truncated much digital participation into relatively superficial engagements within the gated communities such as Facebook, Tiktok, or X in service commercial and commersio-political ends. Some observers assert that digital peer production has crowded out other participatory forms and resulted in a culture of superficiality (Kelty 2020). The incremental design participation in gated digital communities has certainly dampened the vast potential of digital participation forms that were hoped to be achieved in the early 2010s, and, ironically, cut much of the *design* participation that prospered before and alongside the large social media platforms and their algorithmic capitalization of participation. Yet while social media on the whole currently appears to have rather endangered than supported liberal democracies, the import of digital opportunities for design participation is far more variegated: against the crowding-out effects, there has been an explosion of different kinds of participation opportunities and platforms that are effectively used to foster design participation. One can also predict that AI tools may result in similarly varied effects in at once capacitating active citizen action regarding design, planning and innovation yet also undermining and stealing attention from participation and democratic governance.

Design participation is not immune to exploitation and cooptation, nor does it magically produce a power and interest-free space. Kelty (2020) goes as far as to say that participation is always cast between hopes of belonging—being part of, an instance of, a community—and the fears of this experience becoming co-opted or exploited by those in power. Jenssen and Petersen (2016), however, see that participation, in practice, tends to "straddle" this ambiguity by creating an ambivalent middle ground (Hyysalo, Jensen et al. 2016). Indeed, design participation is regularly deployed as a business strategy to gain resources and insights from people for self-interested ends, for instance by the giants of social media (Kelty 2020) and software providers (Pollock et al. 2016). Yet, most of these pursuits seem to be aware of Lincoln's adage of "you can fool some of the people all of the time, and all of the people some of the time, but you cannot fool all of the people all of the time" to the effect that even if the participatory arrangements render most gains to the corporate giant, the participants too are either having some justifiable gain (Pollock and Hyysalo 2014) or do not care about their small contributions becoming monetized (Johnson 2013; von Hippel 2005). Against this backdrop, this book has been cautious about affirming that design participation would somehow "naturally" be a prime vehicle for solving power asymmetries or polarization between actors. As we detailed in Chapters 2, 4, and 5, carefully organized participatory settings can on occasion mitigate the effects of power asymmetries and give access and even empower implicated but excluded actors and result in a narrowing distance between actor positions. But beyond the staged space for participation (Clausen et al., 2020), the processes, experience, or outputs of participation may not further transform the power relations impinging on them (Heiskanen et al. 2010), at least not without further endeavors toward those ends.

The analytical approach to design participation taken in this book is not a recommendation toward paralysis by analysis. Design does require the boldness to act and not just stand by, and most design participation is iterative, which lessens the stakes in taking action. Yet, not considering the contextual requirements or alterative possibilities for how to carry-out design participation has a propensity to result in superficial, ill-suited, or wasteful ways of carrying it out. It is thus crucial to take one's bearings, at least to some extent, to realize what avenues of action are realistically attainable and, thus, for instance, what minimal resources are needed, beyond which design participation cannot be meaningfully pursued.

All the above stated, design participation is a useful avenue, not a panacea for environmental problems. Alone, participation can be just a bandage over an environmental hemorrhage and too slow in the face of climate and biodiversity emergencies. As we underscored in Chapters 1, 4, and 7, the need for actions spans across society and from industry production to consumption and regulation—it is not something that designers alone, or design participation alone, will solve. Yet design participation can importantly support both top-down policy formation and implementation and the refinement, scale-up, and scale-out of bottom-up initiatives. And it can become an important catalyst that bridges the two. In this catalyst role, design participation ought to be pursued for real—for real problems, skillfully, and with implementation in sight—otherwise it may become a vision that acts as a promissory substitute for environmental action on the ground.

Glossary

Alignment work Work that goes into aligning design participation with its contextual conditions; for instance, how design participation action and outputs would relate to the progress of larger planning or political processes.

Black boxing The process where the inner workings of a technology or system become hidden or obscured from outsiders.

Boundary objects Artifacts, tools, or concepts that facilitate communication and collaboration between diverse groups. Boundary objects are plastic enough to adapt to local needs and constraints of the several parties employing them yet robust enough to maintain a common identity across sites.

Codesign A collaborative design approach involving users and stakeholders as design partners; often used as a wider term than participatory design.

Collaborative design work The work that goes into running participation, including facilitation, note-taking, refining participant productions, and keeping things on track.

Constituency building work The work to enlist and involve stakeholders, partners, and co-workers to support design participation efforts.

Co-realization A methodology within extended design-in-use that emphasizes continued design collaboration at the site of use by placing the designer at the user's site to continue the co-evolution of technology and work practices.

Deliberative democracy A participatory process that prioritizes informed opinion formation thorugh discussion, debate, and consensus-building among representatively chosen participants.

Design participation Active engagement of affected and interested parties in shaping design and directing socio-techno-environmental change.

S. Hyysalo, *Design Participation*,
https://doi.org/10.1007/978-3-031-95437-5

Distributed participation Acknowledges that participation processes extend beyond invited events or the initial development context and involve a broader range of actors over time and space and more distributed agency.

Ecology of actors The intertwined network of actors, social worlds, and arenas, some of which are more stable and interlinked and others are more fluid.

Ecology of user knowledge mapping A way to depict an organization and its interconnected set of methods, practices, and sources of information that an organization uses to gain insights about its users.

Extended design-in-use Extends design participation beyond the initial development phase, focusing on user engagement during actual use to refine and adapt designs.

Framing work The work involved in defining the issues for design participation and linking principles, rationales, and concrete proposals for the participation process.

Grassroots innovation networks Localized efforts, such as community-driven renewable energy projects, in which people lead design and innovation efforts with limited input from top-down government or industry programs.

Intermediate designs Designs created during the design process that serve as stepping stones or prototypes toward the final design solution.

Intermediate design work Work that goes into anticipating how the participation might play out and working out/designing all the means, materials, procedures, and arrangements for participation.

Ladder of citizen participation A framework that uses decision control by the participants as a defining point to categorize levels of participation, from superficial tokenism to forms of genuine empowerment.

Lead users People who are ahead of the mainstream regarding a trend and already live the future of others regarding it.

Lead user method A method for identifying and engaging with users who are at the forefront of a trend or have advanced needs, leveraging their insights and expertise to develop innovative products or services.

Material agency The recognition that objects, tools, and technologies actively influence participation and decision-making by enabling or constraining certain actions.

Mid-Range Transition Pathway Toolset (MTPT) Design toolset that facilitates structured discussions about the steps needed to achieve a transition goal.

Multilevel perspective Theory of sociotechnical change that studies it through interactions between three levels: niches (emerging alternatives), regimes (dominant systems), and landscapes (macro-societal trends).

Outcome work The work that goes into turning design participation processes into such shape that produces outputs that participants and outsiders can appropriate and that can have wider outcomes beyond the participation processes.

Participation mix Combining different forms of design participation to cater to diverse participants, different participant needs and preferences for participation, project requirements, and resource constraints.

Participatory design A research and practice approach to design participation where users are considered experts in their own realities and actively and directly participate in the design and decision-making process.

Peer-to-peer design networks Networks that facilitate collaboration and knowledge-sharing among designers, often without a central authority or hierarchy.

Pluriversal design An approach that prioritizes incorporating multiple ontological perspectives in the design process, ensuring all participants are constantly engaging with diverse viewpoints.

Post-mortem evaluations Evaluations conducted after a project or event is completed to reflect on its successes, challenges, and lessons learned.

Producer-user configurations The dynamic relationships between designers (producers) and stakeholders (users) and the predominant ways the interactions, media, and materials move and evolve between them.

Selection work Work of assessing who the potentially affected actors are and how to interest, reach, invite, and select the participants from among them.

Situational matrix A framework for analyzing the constituents of the situations of sociotechnical action, such as design participation.

Tokenism A superficial form of participation where stakeholders are involved only for appearances, without giving them meaningful influence or decision-making power.

Transition arenas Platforms or spaces where stakeholders from different sectors and perspectives come together to discuss, plan, and coordinate actions toward sustainability transitions.

Transition management An approach to steer transitions within a societal sector through envisioning, experimenting, and reflecting on change with a diverse group of front-runners.

User communities Groups of people who share a common interest in a particular product, service, technology, or place and interact and collaborate online or offline to shape it.

References

Abbott, Andrew. 1995. "Things of Boundairies." *Social Research* 62: 857–82.

Abbott, Andrew. 2005. "Linked Ecologies: States and Universities as Environments for Professions." *Sociological Theory* 23 (3): 245–74.

Abel, Bas van, Lucas Evers, Roel Klaassen, and Peter Troxler. 2011. *Open Design Now: Why Design Cannot Remain Exclusive*. San Francisco: BIS Publishers.

Abrell, Thomas, Andreas Benker, and Matti Pihlajamaa. 2018. "User Knowledge Utilization in Innovation of Complex Products and Systems: An Absorptive Capacity Perspective." *Creativity and Innovation Management* 27 (2): 169–82.

Agid, Shana. 2016. "'...It's Your Project, But It's Not Necessarily Your Work...': Infrastructuring, Situatedness, and Designing Relational Practice." In *Proceedings of the 14th Participatory Design Conference: Full Papers—Volume 1*, 81–90. PDC '16. New York, NY, USA: Association for Computing Machinery.

Agid, Shana. 2018. "'Dismantle, Change, Build': Designing Abolition at the Intersections of Local, Large-Scale, and Imagined Infrastructures." *Design Studies*, Participatory Design, 59 (November): 95–116.

Agid, Shana, and Yoko Akama. 2018. "Dance of Designing: Rethinking Position, Relation and Movement in Service Design." *Proceedings of ServDes*.

Aitamurto, Tanja, Dónal Holland, and Sofia Hussain. 2015. "The Open Paradigm in Design Research." *Design Issues* 31 (4): 17–29.

Akera, Atsushi. 2007. "Constructing a Representation for an Ecology of Knowledge: Methodological Advances in the Integration of Knowledge and Its Various Contexts." *Social Studies of Science* 37 (3): 413–41.

Akrich, Madeleine. 1992. "The De-scription of Technical Objects." In *Shaping Technology/Building Society. Studies in Sociotechnical Change*, edited by W. E. Bijker and J. Law, 205–24. Cambridge, MA: MIT Press.

Akrich, Madeleine. 1995. "User Representations: Practices, Methods and Sociology." *Managing Technology in Society. The Approach of Constructive Technology Assessment*, 167–84. London: Cassel Publishers Ltd.

Allen, Robert C. 1983. "Collective Invention." *Journal of Economic Behavior & Organization* 4 (1): 1–24.

Alter, Steven. 2006. *The Work System Method: Connecting People, Processes, and IT for Business Results*. Larkspur: Work System Method.

S. Hyysalo, *Design Participation*,
https://doi.org/10.1007/978-3-031-95437-5

Andriani, Pierpaolo, and Gino Cattani. 2016. "Exaptation as Source of Creativity, Innovation, and Diversity: Introduction to the Special Section." *Industrial and Corporate Change* 25 (1): 115–31.

Antorini, Yun Mi, Albert M. Muñiz Jr., and Tormod Askildsen. 2012. "Collaborating with Customer Communities: Lessons from the LEGO Group." *MIT Sloan Management Review* 53 (3): 73.

Ansell, Christopher, and Jacob Torfing. 2014. *Public Innovation Through Collaboration and Design*. London: Routledge.

Ansell, Christopher, and Jacob Torfing. 2021. *Public Governance as Co-creation: A Strategy for Revitalizing the Public Sector and Rejuvenating Democracy*. Cambridge: Cambridge University Press.

Ardati, Abd Alsattar. 2023. "Ethnographically-Informed Distributed Participatory Design Framework for Sociotechnical Change: Co-designing a Collaborative Training Tool to Support Real-Time Collaborative Writing." PhD Thesis, The University of St. Andrews.

Arnstein, Sherry R. 1969. "A Ladder of Citizen Participation." *Journal of the American Institute of Planners* 35 (4): 216–24.

Asaro, Peter M. 2000. "Transforming Society by Transforming Technology: The Science and Politics of Participatory Design." *Accounting, Management and Information Technologies* 10 (4): 257–90.

Auvinen, Karoliina. 2024. "Accelerating the Energy Transition Toward Zero-Emission District Heating Systems Through Policy Codesign."

Auvinen, Karoliina, Teemu Meriläinen, Laura Saikku, Sampsa Hyysalo, and Jouni K. Juntunen. 2023. "Accelerating Transition Toward District Heating-System Decarbonization by Policy Co-design with Key Investors: Opportunities and Challenges." *Sustainability: Science, Practice and Policy* 19 (1): 2256622.

Bakardjieva, Maria. 2005. "Internet Society: The Internet in Everyday Life." London: Sage.

Bakırlıoğlu, Yekta, and Cindy Kohtala. 2019. "Framing Open Design Through Theoretical Concepts and Practical Applications: A Systematic Literature Review." *Human–Computer Interaction* 34 (5–6): 389–432.

Baldwin, Carliss, and Eric Von Hippel. 2011. "Modeling a Paradigm Shift: From Producer Innovation to User and Open Collaborative Innovation." *Organization Science* 22 (6): 1399–417.

Balka, Kerstin, Christina Raasch, and Cornelius Herstatt. 2009. "Open Source Enters the World of Atoms: A Statistical Analysis of Open Design." *First Monday* 14.

Barcham, Manuhuia. 2022. "Weaving Together a Decolonial Imaginary Through Design for Effective River Management: Pluriversal Ontological Design in Practice." *Design Issues* 38 (1): 5–16.

Bateson, Gregory. 2000. *Steps to an Ecology of Mind: Collected Essays in Anthropology, Psychiatry, Evolution, and Epistemology*. London: University of Chicago press.

Beck, Eevi E. 2002. "P for Political: Participation Is Not Enough." *Scandinavian Journal of Information Systems* 14 (1): 1.

Becker, Howard S. 1982. *Art Worlds*. Berkeley: University of California Press.

Becker, Howard S. 2008. *Tricks of the Trade: How to Think About Your Research While You're Doing It*. London: University of Chicago Press.

Benker, Andreas. 2015. "Design for Usability of Complex Medical Devices: Leading a Technology-Push Innovation Towards User Acceptance." *Journal of Design, Business & Society* 1 (1): 29–42.

Benkler, Yochai. 2006. *The Wealth of Networks: How Social Production Transforms Markets and Freedom*. New Haven: Yale University Press.
Berker, Thomas, Maren Hartmann, Yves Punie, and Katie Ward. 2005. *Domestication of Media and Technology*. UK: McGraw-Hill Education.
Beyer, Hugh, and Karen Holtzblatt. 1998. *Contextual Design: Defining Customer-Centered Systems*. San Diego, USA: Elsevier.
Bherer, Laurence, Mario Gauthier, and Louis Simard. 2017. *The Professionalization of Public Participation*. Vol. 6. Routledge New York.
Bijker, Wiebe E. 1995. *Of Bicycles, Bakelites, and Bulbs: Toward a Theory of Sociotechnical Change*. Cambridge: MIT Press.
Bjerknes, Gro, Pelle Ehn, and Morten Kyng. 1987. *Computers and Democracy—A Scandinavian Challenge*. Gower Publishing.
Björgvinsson, Erling. 2008. "Open-Ended Participatory Design as Prototypical Practice." *CoDesign* 4 (2): 85–99.
Björgvinsson, Erling, and Per-Anders Hillgren. 2009. "Indigenous Design: Healthcare Professional Using Self-Produced Video in Articulating and Developing Work Practices." In *Nordic Design Research Conference 09*. Oslo: NORDES.
Bjögvinsson, Erling, Pelle Ehn, and Per-Anders Hillgren. 2012. "Design Things and Design Thinking: Contemporary Participatory Design Challenges." *Design Issues* 28 (3): 101–16.
Blumer, Herbert. 1969. *Symbolic Interactionism: Perspective and Method*. Berkeley: University of California Press.
Bobrow, Daniel G., and Jack Whalen. 2011. "Community Knowledge Sharing in Practice: The Eureka Story." *Reflections* 4 (2): 47–59.
Bødker, Keld, Finn Kensing, and Jesper Simonsen. 2004. *Participatory IT Design: Designing for Business and Workplace Realities*. Cambridge, MA, and London: MIT Press.
Bødker, Susanne, Christian Dindler, and Ole Sejer Iversen. 2017. "Tying Knots: Participatory Infrastructuring at Work." *Computer Supported Cooperative Work (CSCW)* 26 (1–2): 245–73.
Bogers, Marcel, Allan Afuah, and Bettina Bastian. 2010. "Users as Innovators: A Review, Critique, and Future Research Directions." *Journal of Management* 36 (4): 857–75.
Borning, Alan, Batya Friedman, Janet Davis, and Peyina Lin. 2005. "Informing Public Deliberation: Value Sensitive Design of Indicators for a Large-Scale Urban Simulation." In *ECSCW 2005*, edited by Hans Gellersen, Kjeld Schmidt, Michel Beaudouin-Lafon, and Wendy Mackay, 449–68. Berlin/Heidelberg: Springer-Verlag.
Botero, Andrea. 2013. "Expanding Design Space(s): Design in Communal Endeavours." Doctoral Dissertation, Helsinki, Finland: Aalto University, School of Arts, Design and Architecture.
Botero, Andrea, and Sampsa Hyysalo. 2013. "Ageing Together: Steps Towards Evolutionary Co-design in Everyday Practices." *CoDesign* 9 (1): 37–54.
Botero, Andrea, and Kari-Hans Kommonen. 2009. "Aspects of Social Media Design and Innovation in a Project for Aging Together." *Enhancing Interaction Spaces by Social Media for the Elderly* 21.
Botero, Andrea, and Joanna Saad-Sulonen. 2010. "Enhancing Citizenship: The Role of In-Between Infrastructures." In *Proceedings of the 11th Biennial Participatory Design Conference*, 81–90. Sydney Australia: ACM.

Botero, Andrea, and Joanna Saad-Sulonen. 2022. "Scalar Trajectories in Design: The Case of DIY Cloth Face Masks During the COVID-19 Pandemic." *Artifact* 9 (1): 21.1–21.23.

Botero, Andrea, Iina Oilinki, Kari-Hans Kommonen, and Mariana Salgado. 2002. "Digital Tools for Community Building—Towards Community-Driven Design." In *PDC* (January), 215–19.

Botero, Andrea, Kari-Hans Kommonen, and Sanna Marttila. 2010. "Expanding Design Space: Design-in-Use Activities and Strategies."

Botero, Andrea, Sampsa Hyysalo, Cindy Kohtala, and Jack Whalen. 2020. "Getting Participatory Design Done: From Methods and Choices to Translation Work Across Constituent Domains." *International Journal of Design* 14 (2): 17–34.

Bovaird, Tony. 2007. "Beyond Engagement and Participation: User and Community Coproduction of Public Services." *Public Administration Review* 67 (5): 846–60.

Bovaird, Tony, and Elke Loeffler. 2012. "From Engagement to Co-production: The Contribution of Users and Communities to Outcomes and Public Value." *VOLUNTAS: International Journal of Voluntary and Nonprofit Organizations* 23 (4): 1119–38.

Brown, J. 2002. *The World Café: Shaping Our Futures Through Conversations that Matter*. Whole Systems Associates. Accessed September 15, 2014.

Buchanan, Richard. 2001. "Design Research and the New Learning." *Design Issues* 17 (4): 3–23.

Burrell, Gibson, and Gareth Morgan. 1979. *Sociological Paradigms and Organisational Analysis*. Vol. 248. London: Heinemann.

Büscher, Monika, Satinder Gill, Preben Mogensen, and Dan Shapiro. 2001. "Landscapes of Practice: Bricolage as a Method for Situated Design." *Computer Supported Cooperative Work (CSCW)* 10 (1): 1–28.

Büscher, Monika, Michael Christensen, Klaus Marius Hansen, Preben Mogensen, and Dan Shapiro. 2009. "Bottom-Up, Top-Down? Connecting Software Architecture Design with Use." In *Configuring User-Designer Relations*, edited by Monika Büscher, Roger Slack, Mark Rouncefield, Rob Procter, Mark Hartswood, and Alex Voss, 157–91. London: Springer.

Buur, Jacob, and Kirsten Bagger. 1999. "Replacing Usability Testing with User Dialogue." *Communications of the ACM* 42 (5): 63–66.

Callon, Michel. 1992. "The Dynamics of Techno-Economic Networks." In *Technological Change and Company Strategies*, edited by R. Coombs, P. Saviotti, and V. Walsh, 73–102. London: Academic Press.

Campbell, Colin. 2005. "The Craft Consumer: Culture, Craft and Consumption in a Postmodern Society." *Journal of Consumer Culture* 5 (1): 23–42.

Cardullo, Paolo, and Rob Kitchin. 2019. "Being a 'Citizen' in the Smart City: Up and Down the Scaffold of Smart Citizen Participation in Dublin, Ireland." *GeoJournal* 84 (1): 1–13.

Castellion, George, and Stephen K. Markham. 2013. "Perspective: New Product Failure Rates: Influence of Argumentum ad Populum and Self-Interest." *Journal of Product Innovation Management* 30 (5): 976–79.

Ceschin, Fabrizio, and Idil Gaziulusoy. 2016. "Evolution of Design for Sustainability: From Product Design to Design for System Innovations and Transitions." *Design Studies* 47 (November):118–63.

Ceschin, Fabrizio, and İdil Gaziulusoy. 2019. *Design for Sustainability (Open Access): A Multi-Level Framework from Products to Socio-Technical Systems*. London: Routledge.
Chaiklin, Seth, and Jean Lave. 1993. *Understanding Practice: Perspectives on Activity and Context*. Edited by Roy Pea and John Seely Brown. Cambridge: Cambridge University Press.
Checkland, Peter. 1984. "4—Rethinking a Systems Approach." In *Rethinking the Process of Operational Research & Systems Analysis*, edited by Rolfe Tomlinson and Istavàn Kiss, 43–60. Frontiers of Operational FORS. Amsterdam: Pergamon.
Chilvers, Jason, and Matthew Kearnes, eds. 2015. *Remaking Participation: Science, Environment and Emergent Publics*. London: Routledge.
Chilvers, Jason, and Matthew Kearnes, eds. 2020. "Remaking Participation in Science and Democracy." *Science, Technology, & Human Values* 45 (3): 347–80.
Chilvers, Jason, and Noel Longhurst. 2016. "Participation in Transition(s): Reconceiving Public Engagements in Energy Transitions as Co-produced, Emergent and Diverse." *Journal of Environmental Policy & Planning* 18 (5): 585–607.
Chilvers, Jason, Helen Pallett, and Tom Hargreaves. 2018. "Ecologies of Participation in Socio-Technical Change: The Case of Energy System Transitions." *Energy Research & Social Science* 42 (August): 199–210.
Churchill, Joan, Eric Von Hippel and Mary Sonnack. 2009. *Lead User Project Handbook—A Practical Guide for Lead User Project Teams*. Retrieved October 29, 2011.
Clarke, Adele E. 2005. *Situational Analysis—Grounded Theory After the Postmodern Turn*. Thousand Oaks: Sage.
Clarke, Adele, and Susan Leigh Star. 2003. "Symbolic Interactionist Science, Technology, Information and Biomedicine Studies." In *Handbook of Symbolic Interactionism*, edited by Larry T. Reynolds and Nancy J. Herman, 539–74. Walnut Creek, CA: Alta Mira Press.
Clausen, Christian, Dominique Vinck, Signe Pedersen, and Jens Dorland. 2020. *Staging Collaborative Design and Innovation: An Action-Oriented Participatory Approach*. Cheltenham, UK: Edward Elgar.
Cole, Michael. 1996. *Cultural Psychology: A Once and Future Discipline*. Cambridge: Harvard University Press.
Collins, H. M., and Robert Evans. 2002. "The Third Wave of Science Studies: Studies of Expertise and Experience." *Social Studies of Science* 32 (2): 235–96.
Coombs, Rod, Paolo Saviotti, and Vivien Walsh. 1987. *Economics and Technological Change*. Totowa: Rowman & Littlefield.
Cooney, Richard, Nifeli Stewart, Tania Ivanka, and Neal Haslem. 2018. "Representational Artefacts in Social Problem Solving: A Study from Occupational Rehabilitation." *Design Studies* 56 (May): 149–68.
Cooper, Allan. 2004. *Inmates are Running the Asylum—Why High-Tech Products Drive Us Crazy and How to Restore the Sanity*. IN, USA: Sams Publishing.
Corbett, Steven. 2014. "Challenging the Commodification of Public Spheres: The Hacker Work Ethic in a Free Media Lab." *First Monday* 19.
Crilly, Nathan, David Good, Derek Matravers, and P. John Clarkson. 2008. "Design as Communication: Exploring the Validity and Utility of Relating Intention to Interpretation." *Design Studies* 29 (5): 425–57.
Curedale, Robert. 2013. *Service Design: 250 Essential Methods*. Los Angeles: Design Community College Inc.

Dahlström, Marja, and Sirkka Minkkinen. 2009. *Loppukiri: vaihtoehtoista asumista seniori-iässä*. Juva, Finland: WSOY (Last Spurt: Alternative Living in Senior-Age).

Dalsgaard, Peter. 2012. "Participatory Design in Large-Scale Public Projects: Challenges and Opportunities." *Design Issues* 28 (3): 34–47.

Dalsgaard, Peter, and Eva Eriksson. 2013. "Large-Scale Participation: A Case Study of a Participatory Approach to Developing a New Public Library." In *Proceedings of the SIGCHI Conference on Human Factors in Computing Systems*, 399–408. ACM.

Damodaran, Leela. 1996. "User Involvement in the Systems Design Process-a Practical Guide for Users." *Behaviour & Information Technology* 15 (6): 363–77.

Dant, Tim. 2005. *Materiality and Society*. England: McGraw-Hill International.

David, Paul A. 1985. "Clio and the Economics of QWERTY." *The American Economic Review* 75 (2): 332–37.

David, Paul A. 2007. "Path Dependence: A Foundational Concept for Historical Social Science." *Cliometrica* 1 (2): 91–114.

De Jong, Jeroen P. J., Eric von Hippel, Fred Gault, Jari Kuusisto, and Christina Raasch. 2015. "Market Failure in the Diffusion of Consumer-Developed Innovations: Patterns in Finland." *Research Policy* 44 (10): 1856–65.

Denzin, Norman K. 2017. *The Research Act: A Theoretical Introduction to Sociological Methods*. New York: Routledge.

DeSanctis, Gerardine, and Marshall Scott Poole. 1994. "Capturing the Complexity in Advanced Technology Use: Adaptive Structuration Theory." *Organization Science* 5 (2): 121–47.

Dourish, Paul, and Graham Button. 1998. "On 'Technomethodology': Foundational Relationships Between Ethnomethodology and System Design." *Human–Computer Interaction* 13 (4): 395–432.

Dryzek, John S. 2002. *Deliberative Democracy and Beyond: Liberals, Critics, Contestations*. Oxford: Oxford University Press.

Durrant, Rachael. 2014. "Civil Society Roles in Transition: Towards Sustainable Food?" University of Sussex.

Earthy, Jonathan. 1998. "Usability Maturity Model: Human Centredness Scale." *INUSE Project Deliverable D* 5: 1–34.

Eglash, Ron. 2004. "Appropriating Technology: An Introduction." In *Appropriating Technology: Vernacular Science and Social Power*, edited by Ron Eglash, Jennifer L. Croissant, Giovanna Di Chiro, and Rayvon Fouché, vii–xxi. Minneapolis, MN: University of Minnesota Press.

Ehn, Pelle, and Morten Kyng. 1991. "Cardboard Computers: Mocking-It-Up or Hands-on the Future." In *Design at Work*, 169–95. New Jersey, NJ: Lawrence Erlbaum.

Ehn, Pelle, Elisabet M. Nilsson, and Richard Topgaard. 2014. *Making Futures: Marginal Notes on Innovation, Design, and Democracy*. Cambridge: The MIT Press.

Enell-Nilsson, Mona, Petra Berg, Tatu Marttila, Johan Nyström, Hanna-Kaisa Pernaa, and Arto Rajala. 2019. "Kohti Kestävää Liikkumista Ja Liikennettä 2035: Vaasan Murrosareenan Tulokset" (Towards Sustainable Mobility and Transit 2035: The Results of Vaasa Transition Arena).

Engeström, Yrjö. 1987. *Learning by Expanding. An Activity-Theoretical Approach to Developmental Research*. Helsinki: Orienta-Konsultit Oy.

Engeström, Yrjö. 2000. "Activity Theory as a Framework for Analyzing and Redesigning Work." *Ergonomics* 43 (7): 960–74.

Engeström, Yrjö, and Virginia Escalante. 1996. "Mundane Tool or Object of Affection? The Rise and Fall of the Postal Buddy." In *Context and Consciousness: Activity Theory and Human–Computer Interaction*, edited by Bonnie A. Nardi, 325–74. Cambridge, MA and London: The MIT Press.

Eriksen, Mette Agger. 2009. "Engaging Design Materials, Formats and Framings in Specific, Situated Co-designing: A Micro-material Perspective."

Eriksen, Mette Agger. 2012. *Material Matters in Co-designing: Formatting & Staging with Participating Materials in Co-design Projects, Events & Situations*. Sweden: Malmö University.

Eriksen, Mette Agger, Eva Brandt, Tuuli Mattelmäki, and Kirsikka Vaajakallio. 2014. "Taking Design Games Seriously: Re-connecting Situated Power Relations of People and Materials." In *Proceedings of the 13th Participatory Design Conference: Research Papers-Volume 1*, 101–10. New York, NY: ACM.

Ertner, Marie, Signe Yndigegn, Giacomo Poderi, Joanna Saad-Sulonen, and Helena Karasti. 2023. "Project Engagement Technologies and Their Role in Shaping Conditions for Engagement and Care in Participatory Design Projects." In *Ethical and Methodological Dilemmas in Social Science Interventions*, 19–34. Cham: Springer.

Escobar, Arturo. 2018. *Designs for the Pluriverse: Radical Interdependence, Autonomy, and the Making of Worlds*. Duke University Press.

Eychenne, F. 2012. "Fab Labs Overview. Report, The Fing (Fondation Internet Nouvelle Génération)."

Ferguson, Briony C., Rebekah R. Brown, Niki Frantzeskaki, Fjalar J. de Haan, and Ana Deletic. 2013. "The Enabling Institutional Context for Integrated Water Management: Lessons from Melbourne." *Water Research, Urban Water Management to Increase Sustainability of Cities* 47 (20): 7300–14.

Fischer, Gerhard. 2003. "Meta-Design: Beyond User-Centered and Participatory Design." In *Proceedings of HCI international*. Vol. 4, 88–92.

Fischer, Gerhard. 2007. "Meta-Design: Expanding Boundaries and Redistributing Control in Design." In *Human–Computer Interaction—INTERACT 2007*, 193–206. Berlin, Heidelberg: Springer.

Fischer, Gerhard, and Elisa Giaccardi. 2006. "Meta-Design: A Framework for the Future of End-User Development." In *End User Development*, 427–57. Dordrecht: Springer.

Fischer, Gerhard, and Jonathan Ostwald. 2002. "Seeding, Evolutionary Growth, and Reseeding: Enriching Participatory Design with Informed Participation." In *Engagement and New Technologies: Opening Up the Laboratory*, edited by Neelke Doorn, Daan Schuurbiers, Ibo Van De Poel, and Michael E. Gorman, 16: 55–95. Philosophy of Engineering and Technology. Dordrecht: Springer Netherlands

Fleck, James. 1988. *Innofusion or Diffusation?: The Nature of Technological Development in Robotics*. No. 4. Research Centre for Social Sciences, University of Edinburgh.

Fleck, James. 1993. "Configurations: Crystallizing Contingency." *International Journal of Human Factors in Manufacturing* 3 (1): 15–36.

Flichy, Patrice. 2007a. *The Internet Imaginaire*. Cambridge: MIT Press.

Flichy, Patrice. 2007b. *Understanding Technological Innovation: A Socio-technical Approach*. Cheltenham (UK): Edward Elgar Publishing.

Franke, Nikolaus, Eric Von Hippel, and Martin Schreier. 2006. "Finding Commercially Attractive User Innovations: A Test of Lead-User Theory." *Journal of Product Innovation Management* 23 (4): 301–15.

Frantzeskaki, Niki, Julia Wittmayer, and Derk Loorbach. 2014. "The Role of Partnerships in 'Realising' Urban Sustainability in Rotterdam's City Ports Area, The Netherlands." *Journal of Cleaner Production* 65 (February): 406–17.

Frantzeskaki, Niki, Vanesa Castán Broto, Lars Coenen, and Derk Loorbach. 2017. *Urban Sustainability Transitions*. Vol. 5. New York: Routledge.

Freeman, Stephanie. 2011. "Constructing a Community: Myths and Realities of the Open Development Model." University of Helsinki.

Freeman, Stephanie. 2015. "Immersed in Pellet Technology: Motivation Paths of Innovative DIYers." *Outlines. Critical Practice Studies* 16 (1): 54–80.

Friedman, Batya, and David G. Hendry. 2019. *Value Sensitive Design: Shaping Technology with Moral Imagination*. Cambridge: MIT Press.

Friedman, Batya, Peter Kahn, and Alan Borning. 2002. "Value Sensitive Design: Theory and Methods." University of Washington Technical Report 2 (8).

Friedman, Batya, Peter H. Kahn, Alan Borning, and Alina Huldtgren. 2013. "Value Sensitive Design and Information Systems." In *Early Engagement and New Technologies: Opening Up the Laboratory*, edited by Neelke Doorn, Daan Schuurbiers, Ibo Van De Poel, and Michael E. Gorman, vol. 16, 55–95. Philosophy of Engineering and Technology. Dordrecht, The Netherlands: Springer.

Fry, Tony. 2009. *Design Futuring*, 71–77. Sydney: University of New South Wales Press.

Fuenfschilling, Lea, and Bernhard Truffer. 2014. "The Structuration of Socio-Technical Regimes—Conceptual Foundations from Institutional Theory." *Research Policy* 43 (4): 772–91.

Galabo, Rosendy, Mirian Calvo, Leon Cruickshank, and David Perez Ojeda. 2021. "Improving Engagement Tools to Enable Advocacy in Public Sector Activities." In *Fifteenth International Conference on Design Principles and Practices*.

Gardiner, Paul, and Roy Rothwell. 1985. "Tough Customers: Good Designs." *Design Studies* 6 (1): 7–17.

Gaver, Bill, Tony Dunne, and Elena Pacenti. 1999. "Design: Cultural Probes." *Interactions* 6 (1): 21–29.

Gaziulusoy, A. İdil, and Chris Ryan. 2017a. "Roles of Design in Sustainability Transitions Projects: A Case Study of Visions and Pathways 2040 Project from Australia." *Journal of Cleaner Production* 162: 1297–307.

Gaziulusoy, A. İdil, and Chris Ryan. 2017b. "Shifting Conversations for Sustainability Transitions Using Participatory Design Visioning." *The Design Journal* 20 (Sup1): S1916–26.

Geels, Frank W. 2002. "Technological Transitions as Evolutionary Reconfiguration Processes: A Multi-Level Perspective and a Case-Study." *Research Policy* 31 (8–9): 1257–74.

Geels, Frank W. 2004. "From Sectoral Systems of Innovation to Socio-Technical Systems: Insights About Dynamics and Change from Sociology and Institutional Theory." *Research Policy* 33 (6): 897–920.

Geels, Frank W., and Johan Schot. 2007. "Typology of Sociotechnical Transition Pathways." *Research Policy* 36 (3): 399–417.

Glaser, Vern L., Neil Pollock, and Luciana D'Adderio. 2021. "The Biography of an Algorithm: Performing Algorithmic Technologies in Organizations." *Organization Theory* 2 (2): 26317877211004610.

Glenn, Jerome, and T. J. Gordon, eds. 2009. *Futures Research Methodology Version 3.0*. Washington, DC: The Millennium Project.

Goodman-Deane, Joy, Patrick Langdon, Susannah Clarke, and P. John Clarkson. 2008. "Categorising Design Methods: How Designers View the Roles of User Methods in Design." In *Contemporary Ergonomics*, edited by P. Bust, 273–78. Loughborough: Taylor & Francis.

Gorb, Peter, and Angela Dumas. 1987. "Silent Design." *Design Studies* 8 (3): 150–56.

Gould, John D., and Clayton Lewis. 1985. "Designing for Usability: Key Principles and What Designers Think." *Communications of the ACM* 28 (3): 300–11.

Green, David Philip, Verena Fuchsberger, Nick Taylor, Pernille Bjørn, David Kirk, and Silvia Lindtner. 2019. "Introduction to this Special Issue on Open Design at the Intersection of Making and Manufacturing." *Human–Computer Interaction* 34 (5–6): 379–88.

Greenbaum, Joan, and Morten Kyng. 1991. *Design at Work: Cooperative Design of Computer Systems*. Hillsdale, NJ: Lawrence Erlbaum Associates.

Gregory, Judith. 2000. *Sorcerer's Apprentice: Creating the Electronic Health Record, Re-inventing Medical Records and Patient Care*. San Diego: University of California.

Grönlund, Kimmo, Maija Setälä, and Kaisa Herne. 2010. "Deliberation and Civic Virtue: Lessons from a Citizen Deliberation Experiment." *European Political Science Review* 2 (1): 95–117.

Hagen, Penny, and Toni Robertson. 2010. "Seeding Social Technologies: Strategies for Embedding Design in Use." In *Proceedings of the DRS 2010 Conference. Design and Complexity, Montreal, Canada*.Montreal: DRS.

Hakkarainen, Louna. 2017. "Caring for Technology: Evolving Living Lab Collaboration." Doctoral Dissertation, Helsinki, Finland: Aalto University School of Arts, Design and Architecture, Department of Design.

Hakkarainen, Louna, and Sampsa Hyysalo. 2016. "The Evolution of Intermediary Activities: Broadening the Concept of Facilitation in Living Labs." *Technology Innovation Management Review* 6 (1): 45–58.

Halbinger, Maria A. 2018. "The Role of Makerspaces in Supporting Consumer Innovation and Diffusion: An Empirical Analysis." *Research Policy* 47 (10): 2028–36.

Halskov, Kim, and Peter Dalsgård. 2006. "Inspiration Card Workshops." In *Proceedings of the 6th Conference on Designing Interactive Systems*, 2–11. DIS '06. New York, NY, USA: Association for Computing Machinery.

Hannukainen, Pia, Samuli Mäkinen, and Sampsa Hyysalo. 2017. "Organisational Adoption of the Lead User Method: A Follow-Up Study on Intentions Versus Actions." *International Journal of Business Excellence* 12 (4): 508–36.

Harding, Sandra. 1995. "Just Add Women and Stir?" In *Missing Links: Gender Equity in Science and Technology for Development*, edited by Gender Working Group of UN Commission on Science and Technology for Development. Ottawa, ON, CA: IDRC.

Hartmann, Maren. 2023. *The Routledge Handbook of Media and Technology Domestication*. New York: Routledge.

Hartmann, Mia Rosa Koss, and Rasmus Koss Hartmann. 2023. "Hiding Practices in Employee-User Innovation." *Research Policy* 52 (4): 104728.

Hartswood, Mark, Rob Procter, Roger Slack, Alex Vob, Monika Buscher, Mark Rouncefield, and Philippe Rouchy. 2002. "Co-realisation: Towards a Principled Synthesis of Ethnomethodology and Participatory Design." *Scandinavian Journal of Information Systems* 14 (2): 9–30.

Hasselqvist, Hanna, Mia Hesselgren, and Cristian Bogdan. 2016. "Challenging the Car Norm: Opportunities for ICT to Support Sustainable Transportation Practices."

In *Proceedings of the 2016 CHI Conference on Human Factors in Computing Systems*, 1300–1311. CHI '16. New York, NY, USA: Association for Computing Machinery.

Hassenzahl, Marc, and Noam Tractinsky. 2006. "User Experience—A Research Agenda." *Behaviour & Information Technology* 25 (2): 91–97.

Hasu, Mervi. 2001. "Critical Transition from Developers to Users." Academic Dissertation, Helsinki: University of Helsinki, Department of Education.

Heiskanen, Eva, and Petteri Repo. 2007. "User Involvement and Entrepreneurial Action." *Human Technology* 3 (2): 167–87.

Heiskanen, Eva, Sirkku Kivisaari, Raimo Lovio, and Per Mickwitz. 2009. "Designed to Travel? Transition Management Encounters Environmental and Innovation Policy Histories in Finland." *Policy Sciences* 42 (4): 409–27.

Heiskanen, Eva, Sampsa Hyysalo, Tanja Kotro, and Petteri Repo. 2010. "Constructing Innovative Users and User-Inclusive Innovation Communities." *Technology Analysis & Strategic Management* 22 (4): 495–511.

Henderson, Austin, and Morten Kyng. 1995. "There's No Place Like Home: Continuing Design in Use." In *Readings in Human–Computer Interaction*, edited by Ronald M. Baecker, Jonathan Grudin, William A. S. Buxton, and SAUL Greenberg, 793–803. Interactive Technologies. Morgan Kaufmann.

Hepp, Andreas, and Anne Schmitz. 2022. "The Limits of the Maker Ideology: Local Makerspaces, Experimental Practices, and COVID-19." *Continuum* 36 (2): 199–213.

Hermans, Guido. 2015. "Opening Up Design: Engaging the Layperson in the Design of Everyday Products." Doctoral Dissertation, Umeå, Sweden: Umeå Institute of Design, Faculty of Science and Technology.

Hernberg, Hella. 2022. *Architects as 'Mediators': Socio-Political Roles in Mediating the 'Temporary Use' of Vacant Spaces*. Espoo, Finland: Aalto University.

Hernberg, Hella, and Sampsa Hyysalo. 2024. "Modes of Intermediation: How Intermediaries Engage in Advancing Local Bottom-Up Experimentation." *Environmental Innovation and Societal Transitions* 51 (June): 100849.

Herstatt, Cornelius, and Eric Von Hippel. 1992. "From Experience: Developing New Product Concepts Via the Lead User Method, a Case Study in a 'Low Tech' Field." *Journal of Product Innovation Management* 9: 213–21.

Hesselgren, Mia. 2019. "Designing for Sustainability Practices: RE-DO Design doings, Strategies and Postures." PhD Thesis, KTH Royal Institute of Technology.

Hesselgren, Mia, Hanna Hasselqvist, and Elina Eriksson. 2015. "A Car-Free Year: Providing Vehicles for Change." *Nordes* 1 (6).

Hienerth, Christoph, Eric Von Hippel, and Morten Berg Jensen. 2014. "User Community Vs. Producer Innovation Development Efficiency: A First Empirical Study." *Research Policy* 43 (1): 190–201.

Hillgren, Per-Anders, Anna Seravalli, and Anders Emilson. 2011. "Prototyping and Infrastructuring in Design for Social Innovation." *CoDesign* 7 (3–4): 169–83.

Hodgson, A. M. 1992. "Hexagons for Systems Thinking." *European Journal of Operational Research* 59 (1): 220–30.

Holmström, Helena. 2004. *Community-Based Customer Involvement for Improving Packaged Software Development*. Gothenburg: Göthenburg University.

Holtzblatt, Karen, and Sandra Jones. 1993. "Contextual Inquiry: A Participatory Technique for System Design." In *Participatory Design*, edited by D. Schuler and A. Namioka, 177–210. Hillsdale, NJ: Lawrence Erlbaum Associates.

Holtzblatt, Karen, Jessamyn Burns Wendell, and Shelley Wood. 2005. *Rapid Contextual Design—A How-to Guide to Key Techniques for User-Centered Design*. San Francisco: Morgan Kaufman Publishers.

Hoogma, Remco, René Kemp, Johan Schot, and Bernhard Truffer. 2002. *Experimenting for Sustainable Transport*. Oxford: Taylor & Francis.

Hornyanszky Dalholm, Elisabeth. 1998. *Att Forma Sitt Rum. Fullskalemodellering i Participatoriska Designprocesser* (To Design One's Room. Full Scale Modelin in Participatory Design Processes). Lund: Lunds Tekniska Högskola.

Howells, Jeremy. 2006. "Intermediation and the Role of Intermediaries in Innovation." *Research Policy* 35 (5): 715–28.

Hughes, Everett C. 1993. *The Sociological Eye: Selected Papers*. New Brunswick: Transaction Publishers.

Huybrechts, Liesbeth, Maurizio Teli, Joanna Saad-Sulonen, and Mariacristina Sciannamblo. 2024. "Collectives and Participatory Design." In *Routledge International Handbook of Contemporary Participatory Design*, 51–78. London: Routledge.

Hyödynmaa, Anna. 2016. "Demokratian Leikkikentällä: Osallistava Suunnittelu Hallinnon Ja Demokratian Murroksessa."

Hyvärinen, Jaana, Jung-Joo Lee, and Tuuli Mattelmäki. 2015. "Fragile Liaisons: Challenges in Cross-Organizational Service Networks and the Role of Design." *The Design Journal* 18 (2): 249–68.

Hyysalo, Sampsa. 2010. *Health Technology Development and Use: From Practice-Bound Imagination to Evolving Impacts*. New York: Routledge.

Hyysalo, Sampsa. 2021. *Citizen Activities in Energy Transition: User Innovation, New Communities, and the Shaping of a Sustainable Future*. Taylor & Francis.

Hyysalo, Sampsa, and Louna Hakkarainen. 2014. "What Difference Does a Living Lab Make? Comparing Two Health Technology Innovation Projects." *CoDesign* 10 (3–4): 191–208.

Hyysalo, Sampsa, and Mikael Johnson. 2015. "The User as Relational Entity: Options That Deeper Insight into User Representations Opens for Human-Centered Design." *Information Technology & People* 28 (1): 72–89.

Hyysalo, Sampsa, and Mikael Johnson. 2024. "Making Sense of Methods and Approaches to User Involvement." *The Design Journal* 27 (4): 580–608.

Hyysalo, Sampsa, and Jouni K. Juntunen. 2024. "Series of Configurational Movements: User Activities in Technology Generalization." *Technological Forecasting and Social Change* 200 (March): 123158.

Hyysalo, Sampsa, and Janne Lehenkari. 2002. "Contextualizing Power in Collaborative Design." In *Proceedings of the 7th Biennial Participatory Design Conference*, edited by Thomas Binder, Judith Gregory, and Ina Wagner, 93–104.

Hyysalo, Sampsa, and Janne Lehenkari. 2018. "An Activity-Theoretical Method for Studying User Participation in IS Design." *Methods of Information in Medicine* 42 (February): 398–404.

Hyysalo, Sampsa, and Svetlana Usenyuk. 2015. "The User Dominated Technology Era: Dynamics of Dispersed Peer-Innovation." *Research Policy* 44 (3): 560–76.

Hyysalo, Sampsa, Sebastian Greger, and Zagros Hatami. 2013. "Co-designing the Future Health Care." In *Living+ for Better Living Environments*, 89–93. Aalto ARTS Books.

Hyysalo, Sampsa, Cindy Kohtala, Pia Helminen, Samuli Mäkinen, Virve Miettinen, and Lotta Muurinen. 2014. "Collaborative Futuring with and By Makers." *CoDesign* 10 (3–4): 209–28.

Hyysalo, Sampsa, Petteri Repo, Päivi Timonen, Louna Hakkarainen, and Eva Heiskanen. 2016. "Diversity and Change of User Driven Innovation Modes in Companies." *International Journal of Innovation Management* 20 (2): 1650023.

Hyysalo, Sampsa, Torben Elgaard Jensen, and Nelly Oudshoorn. 2016. *The New Production of Users: Changing Innovation Collectives and Involvement Strategies*. New York: Routledge.

Hyysalo, Sampsa, Tatu Marttila, Sofi Perikangas, and Karoliina Auvinen. 2019. "Codesign for Transitions Governance: A Mid-Range Pathway Creation Toolset for Accelerating Sociotechnical Change." *Design Studies* 63 (July): 181–203.

Hyysalo, Sampsa, Jani Lukkarinen, Paula Kivimaa, Raimo Lovio, Armi Temmes, Mikael Hildén, Tatu Marttila, et al. 2019. "Developing Policy Pathways: Redesigning Transition Arenas for Mid-Range Planning." *Sustainability* 11 (3): 603.

Hyysalo, Sampsa, Sofi Perikangas, Tatu Marttila, and Karoliina Auvinen. 2019. "Intermediate Codesigning in Transitions Governance: Catalysing and Channelling Participant Action." *The Design Journal* 22 (6): 873–94.

Hyysalo, Sampsa, Virve Hyysalo, and Louna Hakkarainen. 2019. "The Work of Democratized Design in Setting-Up a Hosted Citizen-Designer Community." *International Journal of Design* 13 (1): 69–82.

Hyysalo, Sampsa, Eva Heiskanen, Jani Lukkarinen, Kaisa Matschoss, Mikko Jalas, Paula Kivimaa, Jouni K. Juntunen, Fanni Moilanen, Pekka Murto, and Eeva Primmer. 2022. "Market Intermediation and Its Embeddedness—Lessons from the Finnish Energy Transition." *Environmental Innovation and Societal Transitions* 42 (March): 184–200.

Hyysalo, Sampsa, Satu Lähteenoja, Tatu Marttila, Kaisa Savolainen, Ari-Pekka Auvinen, Jani Jani Lukkarinen, Juuli Närhi, Joona Lehtomäki, Kirsi-Marja Lonkila, and Kristiina Niikkonen. 2022. "Ekologinen Siirtymä Luonnon Monimuotoisuuspolitiikassa: Biodiversiteettiareenan 2021 Tulokset."

Hyysalo, Virve. 2021. "Ala Laputtaa!–Kevyt Ja Helposti Monistettava Menetelmä Osana Oodin Yhteiskehittämistä." In *Miettinen, Satu (Toim.): Muotoilun Avaimet Älykkääseen Teollisuuteen Ja Liiketoiminnan Ketterään Kehittämiseen* [Start Placing the Stickies!—Light and Easily Replicable Method as Part of Oodi's Codesign]. Helsinki: Teknologiainfo Teknova Oy.

Hyysalo, Virve. 2022. "Yhteissuunnittelu Ja Palvelumuotoilu Julkisen Sektorin Kehityssuuntina: Kaupunkilaiset Tekijöinä, Käyttäjinä Ja Kumppaneina Keskustakirjasto Oodin Suunnittelussa" [Codesign and Service Design as Development Directions for the Public Sector: Citizens as Makers, Users and Partners in the Planning of Central Library Oodi]. Rovaniemi: University of Lapland.

Hyysalo, Virve, and Sampsa Hyysalo. 2018. "The Mundane and Strategic Work in Collaborative Design." *Design Issues* 34 (3): 42–58.

IAP. 2023. IAP2 Spectrum for Public Participation by International Association of Participation. https://www.iap2.org/page/pillars.

Iivari, Juhani, Rudy Hirschheim, and Heinz K. Klein. 2000. "A Dynamic Framework for Classifying Information Systems Development Methodologies and Approaches." *Journal of Management Information Systems* 17 (3): 179–218.

International Energy Agency (IEA). 2023. *World Energy Outlook 2023*. Paris: OECD Publishing.

The Intergovernmental Panel on Climate Change (IPCC). 2018. Global Warming of 1.5°C. An IPCC Special Report on the Impacts of Global Warming of 1.5°C Above Pre-industrial Levels and Related Global Greenhouse Gas Emission Pathways, in the

Context of Strengthening the Global Response to the Threat of Climate Change, Sustainable Development, and Efforts to Eradicate Poverty [V. Masson-Delmotte, P. Zhai, H. O. Pörtner, D. Roberts, J. Skea, P.R. Shukla, A. Pirani, W. Moufouma-Okia, C. Péan, R. Pidcock, S. Connors, J. B. R. Matthews, Y. Chen, X. Zhou, M. I. Gomis, E. Lonnoy, T. Maycock, M. Tignor, and T. Waterfield (eds.)]. In Press.

IPBES. 2019. "Global assessment report on biodiversity and ecosystem services of the Intergovernmental Science-Policy Platform on Biodiversity and Ecosystem Services." Edited by E. S. Brondizio, J. Settele, S. Díaz, and H. T. Ngo, 1148 pages. IPBES Secretariat, Bonn, Germany.

Irwin, Terry. 2015. "Transition Design: A Proposal for a New Area of Design Practice, Study, and Research." *Design and Culture* 7 (2): 229–46.

Irwin, Alan, Torben Elgaard Jensen, and Kevin E. Jones. 2013. "The Good, the Bad and the Perfect: Criticizing Engagement Practice." *Social Studies of Science* 43 (1): 118–35.

Irwin, Terry, Gideon Kossoff, Cameron Tonkinwise, and Peter Scupelli. 2015. *Transition Design 2015: A New Area of Design Research, Practice and Study that Proposes Design-Led Societal Transition Toward More Sustainable Futures*. Pittsburgh, PA: Carnegie Mellon University.

Jacobs, Claus D., Chris Steyaert, and Florian Überbacher. 2013. "Anticipating Intended Users: Prospective Sensemaking in Technology Development." *Technology Analysis & Strategic Management* 25 (9): 1027–43.

Jacobsen, Niels Ebbe, Morten Hertzum, and Bonnie E. John. 1998. "The Evaluator Effect in Usability Tests." In *CHI 98 Conference Summary on Human Factors in Computing Systems*, 255–56.

Jalas, Mikko, Sampsa Hyysalo, Eva Heiskanen, Raimo Lovio, Ari Nissinen, Maija Mattinen, Jenny Rinkinen, Jouni K. Juntunen, Pasi Tainio, and Heli Nissilä. 2017. "Everyday Experimentation in Energy Transition: A Practice-Theoretical View." *Journal of Cleaner Production*, Experimentation for Climate Change Solutions, 169 (December): 77–84.

Jasanoff, Sheila, and Sang-Hyun Kim, eds. 2015. *Dreamscapes of Modernity: Sociotechnical Imaginaries and the Fabrication of Power*. University of Chicago Press.

Jeffries, R., D. Wixon, E. L.-C. Law, E. Hvannberg, and G. Cockton. 2007. *Maturing Usability: Quality in Software, Interaction and Value*. Springer Science & Business Media.

Jegou, F., and Ezio Manzini. 2008. *Collaborative Services. Social Innovation and Design for Sustainability*. Milan: Edizioni POLI.design.

Jensen, Torben Elgaard. 2012. "Intervention by Invitation: New Concerns and New Versions of the User in STS." *Science & Technology Studies* 25 (1): 13–36.

Jensen, Torben Elgaard, and Morten Krogh Petersen. 2016. "Straddling, Betting and Passing: The Configuration of User Involvement in Cross-sectorial Innovation Projects." In *The New Production of Users*. Routledge.

Jeppesen, Lars Bo, and Måns J. Molin. 2003. "Consumers as Co-developers: Learning and Innovation Outside the Firm." *Technology Analysis & Strategic Management* 15 (3): 363–83.

Jeppesen, Sandra, Anna Kruzynski, Rachel Sarrasin, and Emilie Breton. 2014. "The Anarchist Commons." *Ephemera: Theory and Politics in Organisation* 14 (4): 879–900.

Johnson, Jim. 1988. "Mixing Humans and Nonhumans Together: The Sociology of a Door-Closer." *Social Problems* 35 (3): 298–310.

Johnson, Mikael. 2013. *How Social Media Changes User-Centred Design—Cumulative and Strategic User Involvement with Respect to Developer—User Social Distance*. Aalto University.

Johnson, Mikael, Sampsa Hyysalo, and Sakari Tamminen. 2010. "The Virtuality of Virtual Worlds, or What We Can Learn from Playacting Horse Girls and Marginalized Developers." *Symbolic Interaction* 33 (4): 603–33.

Johnson, Mikael, Hajar Mozaffar, Gian Marco Campagnolo, Sampsa Hyysalo, Neil Pollock, and Robin Williams. 2014. "The Managed Prosumer: Evolving Knowledge Strategies in the Design of Information Infrastructures." *Information, Communication & Society* 17 (7): 795–813.

Johnson, Mikael, Sampsa Hyysalo, Samuli Mäkinen, Pia Helminen, Kaisa Savolainen, and Louna Hakkarainen. 2014. "From Recipes to Meals... and Dietary Regimes: Method Mixes as Key Emerging Topic in Human-Centred Design." In *Proceedings of the 8th Nordic Conference on Human–Computer Interaction: Fun, Fast, Foundational*, 343–52. NordiCHI '14. New York, NY, USA: Association for Computing Machinery.

Junginger, Sabine. 2008. "Product Development as a Vehicle for Organizational Change." *Design Issues* 24 (1): 26–35.

Juntunen, Jouni K. 2014. "Prosuming Energy—User Innovation and New Energy Communities in Renewable Micro-generation." Doctoral Dissertation 142/2014, Helsinki, Finland: Aalto University School of Business, Department of Management Studies.

Kanger, Laur, Frank W. Geels, Benjamin Sovacool, and Johan Schot. 2019. "Technological Diffusion as a Process of Societal Embedding: Lessons from Historical Automobile Transitions for Future Electric Mobility." *Transportation Research Part D: Transport and Environment* 71: 47–66.

Karasti, Helena. 2014. "Infrastructuring in Participatory Design." In *Proceedings of the 13th Participatory Design Conference: Research Papers—Volume 1*, 141–50. PDC '14. New York, NY, USA: Association for Computing Machinery.

Karasti, Helena, and Anna-Liisa Syrjänen. 2004. "Artful Infrastructuring in Two Cases of Community PD." In *Proceedings of the Eighth Conference on Participatory Design: Artful Integration: Interweaving Media, Materials and Practices-Volume 1*, 20–30. ACM.

Karasti, Helena, Karen S. Baker, and Florence Millerand. 2010. "Infrastructure Time: Long-Term Matters in Collaborative Development." *Computer Supported Cooperative Work (CSCW)* 19 (3): 377–415.

Karasti, Helena, Andrea Botero, Elena Parmiggiani, Karen Baker, Sanna Marttila, Joanna Saad-Sulonen, and Hanne Cecilie Geirbo. 2018. "Infrastructuring in PD: What Does Infrastructuring Look Like? When Does It Look Like that?" In *Proceedings of the 15th Participatory Design Conference: Short Papers, Situated Actions, Workshops and Tutorial—Volume 2*, 1–3. Hasselt and Genk Belgium: ACM.

Kaulio, Matti A. 1998. "Customer, Consumer and User Involvement in Product Development: A Framework and a Review of Selected Methods." *Total Quality Management* 9 (1): 141–49.

Keinonen, Turkka K., Vesa Jääskö, and Tuuli M. Mattelmäki. 2008. "Three-in-One User Study for Focused Collaboration." *International Journal of Design (Online)* 2 (1): 1–10.

Kelty, Christopher M. 2020. *The Participant: A Century of Participation in Four Stories*. University of Chicago Press.

Kemp, René, Johan Schot, and Remco Hoogma. 1998. "Regime Shifts to Sustainability Through Processes of Niche Formation: The Approach of Strategic Niche Management." *Technology Analysis & Strategic Management* 10 (2): 175–98.

Kemp, René, Derk Loorbach, and Jan Rotmans. 2007. "Transition Management as a Model for Managing Processes of Co-evolution Towards Sustainable Development." *International Journal of Sustainable Development & World Ecology* 14 (1): 78–91.

Kensing, Finn, and Halskov Madsen. 1991. *Generating Visions: Future Workshops and Metaphorical*. Edited by Joan Greenbaum and Morten Kyng, 155–68. Hillsdale, NJ: Lawrence Erlbaum Associates.

Kensing, Finn, Jesper Simonsen, and Keld Bodker. 1998. "MUST: A Method for Participatory Design." *Human–Computer Interaction* 13 (2): 167–98.

Kimbell, Lucy, and Guy Julier. 2019. "Confronting Bureaucracies and Assessing Value in the Co-production of Social Design Research." *CoDesign* 15 (1): 8–23.

Kivelä, Kaisa, Sanna Tyyri-Pohjonen, and Krista Kinnunen. 2013. *Living+: For Better Living Environments*. Espoo: Aalto ARTS Books.

Kivimaa, Paula, and Florian Kern. 2016. "Creative Destruction or Mere Niche Support? Innovation Policy Mixes for Sustainability Transitions." *Research Policy* 45 (1): 205–17.

Kivimaa, Paula, Wouter Boon, Sampsa Hyysalo, and Laurens Klerkx. 2018. "Towards a Typology of Intermediaries in Sustainability Transitions: A Systematic Review and a Research Agenda." *Research Policy*.

Kivimaa, Paula, Sampsa Hyysalo, Wouter Boon, Laurens Klerkx, Mari Martiskainen, and Johan Schot. 2019. "Passing the Baton: How Intermediaries Advance Sustainability Transitions in Different Phases." *Environmental Innovation and Societal Transitions* 31: 110–25.

Kivisaari, Sirkku, Raimo Lovio, and Erja Väyrynen. 2004. "Chapter 10: Managing Experiments for Transition: Examples of Societal Embedding in Energy and Health Care Sectors." In *System Innovation and the Transition to Sustainability*, edited by B. Elzen, F. W. Geels, and K. Green. Cheltenham, UK: Edward Elgar Publishing.

Knott, Stephen. 2013. "Design in the Age of Prosumption: The Craft of Design After the Object." *Design and Culture* 5 (1): 45–67.

Kohler, Jonathan, Frank Geels, Florian Kern, Jochen Markard, Anna Wieczorek, Floortje Alkemade, Flor Avelion, et al. 2019. "An Agenda for Sustainability Transitions Research: State of the Art and Future Directions." *Environmental Innovation and Societal Transitions* 31 (January): 1–32.

Kohtala, Cindy. 2016. *Making Sustainability : How Fab Labs Address Environmental Issues*. Aalto University.

Kohtala, Cindy. 2017. "Making 'Making' Critical: How Sustainability is Constituted in Fab Lab Ideology." *The Design Journal* 20 (3): 375–94.

Kohtala, Cindy. 2018. "The Sociomateriality of FabLabs: Configurations of a Printing Service or Counter-Context?" *The Journal of Peer Production* 1 (12): 92–110.

Kohtala, Cindy, and Camille Bosqué. 2014. "The Story of MIT-Fablab Norway: Community Embedding of Peer Production." *Journal of Peer Production* 5: 8.

Kohtala, Cindy, and Sampsa Hyysalo. 2015. "Anticipated Environmental Sustainability of Personal Fabrication." *Journal of Cleaner Production* 99 (July): 333–44.

Kohtala, Cindy, Sampsa Hyysalo, and Jack Whalen. 2020. "A Taxonomy of Users' Active Design Engagement in the 21st Century." *Design Studies* 67 (March): 27–54.

Konrad, Kornelia. 2008. "Dynamics of Type-Based Scenarios of Use: Opening Processes in Early Phases of Interactive Television and Electronic Marketplaces." *Science & Technology Studies* 21 (2): 3–26.

Kopytoff, Igor. 1986. "The Cultural Biography of Things: Commoditization as Process." In *The Social Life of Things: Commodities in Cultural Perspective*, edited by Arjun Appadurai, 64–94. Cambridge: Cambridge University Press.

Kotro, Tanja. 2005. "Hobbyist Knowing in Product Development: Desirable Objects and Passion for Sports in Suunto Corporation." Academic Dissertation, Helsinki: National Consumer Research Center and University of Art and Design Helsinki.

van Kuijk, Jasper, and R. Staats. 2012. "Design for Usability: Methods & Tools: A Practitioner's Guide." IOP IPCR Design for Usability Research Project.

Kujala, Sari. 2003. "User Involvement: A Review of the Benefits and Challenges." *Behaviour & Information Technology* 22 (1): 1–16.

Kuznetsov, Stacey, and Eric Paulos. 2010. "Rise of the Expert Amateur: DIY Projects, Communities, and Cultures." In *Proceedings of the 6th Nordic Conference on Human–Computer Interaction: Extending Boundaries*, 295–304. NordiCHI '10. New York, NY, USA: Association for Computing Machinery.

Lähteenoja, Satu. 2024. "Advancing Sustainability Transformations-Co-Design for Sustainable Development Policies."

Lähteenoja, Satu, Sampsa Hyysalo, and Tatu Marttila. 2022. "Kuusi Muutospolkua Agenda2030-Tavoitteisiin. Agenda2030-Murrosareenatyön Tulokset." In *Kestävän Kehityksen Toimikunnan Agenda2030-Tiekartta*, 130–78. Valtioneuvoston kanslia (Six Transition Pathways to Agenda 2030 Goals. The Results of Agenda 2030 Transition Arena Work).

Lähteenoja, Satu, Sampsa Hyysalo, Jani Lukkarinen, Tatu Marttila, Heli Saarikoski, Maija Faehnle, and Lasse Peltonen. 2022. "What Does It Take to Study Learning in Transitions? A Case of Citizen Energy in Finland." *Sustainability: Science, Practice and Policy* 18 (1): 651–64.

Lähteenoja, Satu, Tatu Marttila, İdil Gaziulusoy, and Sampsa Hyysalo. 2023. "Transition Co-design Dynamics in High Level Policy Processes." *Design Studies* 88 (September): 101207.

Lamb, Roberta, and Rob Kling. 2003. "Reconceptualizing Users as Social Actors in Information Systems Research." *MIS Quarterly* 27 (2): 197–236.

Latour, Bruno. 1987. *Science in Action: How to Follow Scientists and Engineers Through Society*. Cambridge: Harvard University Press.

Latour, Bruno. 1993. *On Technical Mediation. The Messenger Lectures on the Evolution of Civilization, Cornell University*. Lund: University of Lund.

Latour, Bruno. 1996a. "On Interobjectivity." *Mind, Culture, and Activity* 3 (4): 228–45.

Latour, Bruno. 1996b. "Pursuing the Discussion of Interobjectivity with a Few Friends." *Mind, Culture, and Activity* 3 (4): 266–69.

Latour, Bruno. 1999. *Pandora's Hope: Essays on the Reality of Science Studies*. Cambridge: Harvard University Press.

Latour, Bruno. 2005. *Reassembling the Social: An Introduction to Actor-Network-Theory*. Oxford, UK: Oxford University Press.

Lave, Jean, and Etienne Wenger. 1991. *Situated Learning: Legitimate Peripheral Participation*. Cambridge: Cambridge University Press.

Lave, Jean, Roy Pea, and John Seely Brown. 1993. "The Practice of Learning." In *Understanding Practice: Perspectives on Activity and Context*, edited by Seth Chaiklin and Jean Lave, 3–32. Cambridge: Cambridge University Press.

Law, Effie Lai-Chong, Virpi Roto, Marc Hassenzahl, Arnold POS Vermeeren, and Joke Kort. 2009. "Understanding, Scoping and Defining User Experience: A Survey Approach." In *Proceedings of the SIGCHI Conference on Human Factors in Computing Systems*, 719–28. ACM.

Lee, Caroline W. 2014. *Do-It-Yourself Democracy: The Rise of the Public Engagement Industry*. New York: Oxford University Press.

Lee, Jung-Joo, Miia Jaatinen, Anna Salmi, Tuuli Mattelmäki, Riitta Smeds, and Mari Holopainen. 2018. "Design Choices Framework for Co-creation Projects." *International Journal of Design* 12 (2): 15–31.

Lehtimäki, Kirsti, and Taina Rajanti. 2007. "Local Voice in a Global World—User-Centered Design in Support of Everyday Practices." In *Universal Access in Human Computer Interaction. Coping with Diversity*, 197–206. Berlin, Heidelberg: Springer.

Leminen, Seppo. 2015. "Living Labs as Open Innovation Networks-Networks, Roles and Innovation Outcomes." PhD Thesis, Espoo: Aalto University.

Leminen, Seppo, and Mika Westerlund. 2012. "Towards Innovation in Living Labs Networks." *International Journal of Product Development* 17 (1): 43–59.

Leonard-Barton, Dorothy. 1995. *Wellsprings of Knowledge: Building and Sustaining the Sources of Innovation*. Boston, MA: Harvard Business School Press.

Lettenmeier, Michael. 2018. *A Sustainable Level of Material Footprint—Benchmark for Designing One-Planet Lifestyles*. Finland: Aalto University.

Lieberman, Henry, Fabio Paternò, Markus Klann, and Volker Wulf. 2006. "End-User Development: An Emerging Paradigm." In *End User Development*, 1–8. Dordrecht, The Netherlands: Springer.

Liikkanen, Lassi A., and Antti Salovaara. 2015. "Music on YouTube: User Engagement with Traditional, User-Appropriated and Derivative Videos." *Computers in Human Behavior* 50: 108–24.

Lockton, Dan. 2017. "Design with Intent and the Field of Design for Sustainable Behaviour." In *Living Labs: Design and Assessment of Sustainable*, edited by D. Keyson, O. Guerra-Santin, and D. Lockton, 75–88. Berlin, Heidelberg: Springer.

Loorbach, Derk, and Jan Rotmans. 2006. "Managing Transitions for Sustainable Development." In *Understanding Industrial Transformation: Environment & Policy*, edited by X. Olsthoorn, and A. Wieczorek, vol. 44, 187–206. Dordrecht: Springer.

Loorbach, Derk, and Jan Rotmans. 2010. "The Practice of Transition Management: Examples and Lessons from Four Distinct Cases." *Futures* 42 (3): 237–46.

Lukkarinen, Jani, Tatu Marttila, Heli Saarikoski, Karoliina Auvinen, Maija Faehnle, Sampsa Hyysalo, Hanna-Liisa Kangas, Satu Lähteenoja, Lasse Peltonen, and Miikka Salo. 2020. "Taloyhtiöistä Tulevaisuuden Energiatuottajia: Muutospolut Vuoteen 2035 Ja Murrosareena Tiedon Yhteistuotannon Menetelmänä" (Turning Housing Companies into Future Energy Producers: Transition Pathways to Year 2035 and Transition Arena as a Method for Knowledge Co-production).

Lukkarinen, Jani, Fanny Groundstroem, David Lazarevic, Satu Lähteenoja, and Kati Pitkänen. 2023. "Yhteisöt Ja Suunnittelu Haja-Asutusalueiden Ilmastoviisauden Avaimina. Joensuun Maaseutuareenan Muutospolut Ja Viestit" (Communities and Planning as Keys to Rural Area Climate Wisdom: Pathways and Messages of Rural Transition Arena).

Lukkarinen, Jani, Maija Faehnle, Fanny Groundstroem, Hanna-Liisa Kangas, David Lazarevic, Satu Lähteenoja, Niina Pykäläinen, Joonas Salmijärvi, and Kaisa Savolainen. 2023a. "Arki Ja Politiikka Tukemaan Ilmastoviisautta Lähiöissä. Mellunkylän Lähiöareenan Muutospolut Ja Viestit" (Policy to Support Climate Wise Living in the Suburbs: The Transition Pathways and Messages from the Suburb Transition Arena in Mellunkylä).

Lukkarinen, Jani, Miikka Salo, Maija Faehnle, Heli Saarikoski, Sampsa Hyysalo, Karoliina Auvinen, Satu Lähteenoja, and Tatu Marttila. 2023b. "Citizen Energy Lost in Sustainability Transitions: Knowledge Co-production in a Complex Governance Context." *Energy Research & Social Science* 96: 102932.

Lundvall, Bengt-Ake. 1985. "Product Innovation and User-Producer Interaction." *The Learning Economy and the Economics of Hope* 19 (2): 19–60.

Lüthje, Christian, and Cornelius Herstatt. 2004. "The Lead User Method: An Outline of Empirical Findings and Issues for Future Research." *R&D Management* 34 (5): 553–568.

MacKenzie, Donald, and Judy Wajcman. 1999. *The Social Shaping of Technology*. Buckingham, UK: Open University.

Maguire, Martin. 2001. "Methods to Support Human-Centred Design." *International Journal of Human–Computer Studies* 55 (4): 587–634.

Mäkinen, Samuli, Sampsa Hyysalo, and Mikael Johnson. 2018. "Ecologies of User Knowledge: Linking User Insight in Organisations to Specific Projects." *Technology Analysis & Strategic Management* 31 (3): 340–55.

Manzini, Ezio. 2014. "Making Things Happen: Social Innovation and Design." *Design Issues* 30 (1): 57–66.

Manzini, Ezio. 2015. *Design, When Everybody Designs: An Introduction to Design for Social Innovation*. Cambridge: MIT Press.

Manzini, Ezio, and Francesca Rizzo. 2011. "Small Projects/Large Changes: Participatory Design as an Open Participated Process." *CoDesign* 7 (3–4): 199–215.

Margolin, Victor. 1997. "Getting to Know the User." *Design Studies* 18 (3): 227–36.

Markard, Jochen. 2018. "The Life Cycle of Technological Innovation Systems." *Technological Forecasting and Social Change*.

Marres, Noortje. 2012. *Material Participation: Technology, the Environment and Everyday Publics*. Palgrave Macmillan; 1st ed. 2012 edition.

Marttila, Sanna, Andrea Botero, and Joanna Saad-Sulonen. 2014. "Towards Commons Design in Participatory Design." In *Proceedings of the 13th Participatory Design Conference: Short Papers, Industry Cases, Workshop Descriptions, Doctoral Consortium Papers, and Keynote Abstracts—Volume 2*, 9–12. PDC '14. New York, NY, USA: Association for Computing Machinery.

Marttila, Sanna, Elisabet M. Nilsson, and Anna Seravalli. 2014. "Opening Production: Design and Commons." In *Making Futures: Marginal Notes on Innovation, Design, and Democracy*, edited by Pelle Ehn, Elisabet M. Nilsson, and Richard Topgaard. The MIT Press.

Marttila, Tatu, Jani Lukkarinen, Sampsa Hyysalo, David Lazarevic, and Helena Valve. 2023. "Mid-range Transition Arenas as Catalysts in a Circular Economy." In *The Routledge Handbook of Catalysts for a Sustainable Circular Economy*, 339–58. Routledge.

Mattelmäki, Tuuli. 2006. *Design Probes*. Helsinki: University of Art and Design Helsinki.

Mattelmäki, Tuuli, Eva Brandt, and Kirsikka Vaajakallio. 2011. "On Designing Open-Ended Interpretations for Collaborative Design Exploration." *CoDesign* 7 (2): 79–93.
McLaughlin, Janice, Paul Rosen, David Skinner, and Andrew Webster. 2002. *Valuing Technology: Organisations, Culture and Change*. London: Routledge.
Meelen, Toon, Bernhard Truffer, and Tim Schwanen. 2019. "Virtual User Communities Contributing to Upscaling Innovations in Transitions: The Case of Electric Vehicles." *Environmental Innovation and Societal Transitions* 31: 96–109.
Miettinen, Reijo, and Mervi Hasu. 2002. "Articulating User-Needs in Collaborative Design: Towards Activity-Theoretical Approach." *Computer Supported Collaborative Work* 11: 129–51.
Miettinen, Virve. 2018. "Redefining the Library: Co-designing for Our Future Selves and Cities." *Public Library Quarterly* 37 (1): 8–20.
Miller, Clark A., and Carina Wyborn. 2020. "Co-Production in Global Sustainability: Histories and Theories." *Environmental Science & Policy*, 113: 88–95.
Miller, Daniel, and Don Slater. 2007. "Moments and Movements in the Study of Consumer Culture: A Discussion Between Daniel Miller and Don Slater." *Journal of Consumer Culture* 7 (5): 5–23.
Mok, Luisa, and Sampsa Hyysalo. 2017. "Designing for Energy Transition Through Value Sensitive Design." *Design Studies* 54: 162–183.
Mol, Annemarie. 2002. *The Body Multiple: Ontology in Medical Practice*. USA: Duke University Press.
Mosconi, Gaia, Matthias Korn, Christian Reuter, Peter Tolmie, Maurizio Teli, and Volkmar Pipek. 2017. "From Facebook to the Neighbourhood: Infrastructuring of Hybrid Community Engagement." *Computer Supported Cooperative Work (CSCW)* 26 (4): 959–1003.
Mozaffar, Hajar. 2016. "User Communities as Multifunctional Spaces: Innovation, Collective Voice, Demand Articulation, Peer Informing and Professional Identity (and More)." In *The New Production of Users: Changing Innovation Collectives and Involvement Strategies*, edited by S. Hyysalo, T. Jensen, and N. Oudshoorn, 219–246. Routledge.
Muller, Michael, and Allison Druin. 2012. "Participatory Design: The Third Space in Human–Computer Interaction." In *Human Computer Interaction Handbook*, 3rd ed. CRC Press.
Muller, Michael J., and Sarah Kuhn. 1993. "Participatory Design." *Communications of the ACM* 36 (6): 24–28.
Muller, Michael J., Daniel M. Wildman, and Ellen A. White. 1993. "Taxonomy of PD Practices: A Brief Practitioner's Guide." *Communications of the ACM* 36 (4): 26–28.
Murto, Pekka, Mikko Jalas, Jouni Juntunen, and Sampsa Hyysalo. 2019a. "Devices and Strategies: An Analysis of Managing Complexity in Energy Retrofit Projects." *Renewable and Sustainable Energy Reviews* 114: 109294.
Murto, Pekka, Mikko Jalas, Jouni Juntunen, and Sampsa Hyysalo. 2019b. "The Difficult Process of Adopting a Comprehensive Energy Retrofit in Housing Companies: Barriers Posed by Nascent Markets and Complicated Calculability." *Energy Policy* 132 (September): 955–64.
Naito, Ena, and Andrea Botero. 2024. "Growing Conditions for Growing: Infrastructures for Designing with Living Organisms in Fabrication Settings." In *Proceedings*

of the 13th Nordic Conference on Human–Computer Interaction, 1–14. NordiCHI '24. New York, NY, USA: Association for Computing Machinery.
Nardi, Bonnie A. 1996. *Context and Consciousness: Activity Theory and Human–Computer Interaction*. Cambridge, MA: MIT Press.
Nascimento, Susana, and Alexandre Pólvora. 2013. "Opening Up Technologies to the Social: Between Interdisciplinarity and Citizen Participation." *Design Issues* 29 (4): 31–40.
Neale, Wendy, and Craig Hobern. 2017. "Innovation Begins with Inclusion: Integrating the Fab Charter into Every Day at Fab Lab Wgtn." In *Proceedings of the Fab13 Research Stream*. Paper Presented at Fab13, Santiago de Chile.
Nielsen, Jakob. 2017. "A 100-Year View of User Experience." Nielsen Norman Group. December 24, 2017.
Norman, Donald A., and Stephen W. Draper. 1986. *User Centered System Design; New Perspectives on Human–Computer Interaction*. USA: L. Erlbaum Associates Inc.
Numminen, Sini, Marika Silvikko de Villafranca, and Sampsa Hyysalo. 2024. "Hybridization and Accumulation of Space-Heating Systems in Finnish Detached Housing." *Energy Strategy Reviews* 54: 101435.
Okamura, Kazuo, Wanda J. Orlikowski, Masayo Fujimoto, and JoAnne Yates. 1994. "Helping CSCW Applications Succeed: The Role of Mediators in the Context of Use." In *Proceedings of the 1994 ACM Conference on Computer Supported Cooperative Work*, 55–65. CSCW '94. New York, NY, USA: Association for Computing Machinery.
Olsson, Eva. 2004. "What Active Users and Designers Contribute in the Design Process." *Interacting with Computers* 16 (2): 377–401.
van Oost, Ellen, Stefan Verhaegh, and Nelly Oudshoorn. 2009. "From Innovation Community to Community Innovation: User-Initiated Innovation in Wireless Leiden." *Science, Technology, & Human Values* 34 (2): 182–205.
Orlikowski, Wanda J. 2000. "Using Technology and Constituting Structures: A Practice Lens for Studying Technology in Organizations." *Organization Science* 11 (4): 404–28.
Oudshoorn, Nelly, and Trevor Pinch. 2003. *How Users Matter?: The Co-construction of Users and Technologies*. Cambridge, MA: MIT Press.
Oudshoorn, Nelly, Els Rommes, and Marcelle Stienstra. 2004. "Configuring the User as Everybody: Gender and Design in Information and Communication Technologies." *Science, Technology & Human Values* 29 (1): 30–63.
Özkil, Ali Gürcan. 2017. "Collective Design in 3D Printing: A Large Scale Empirical Study of Designs, Designers and Evolution." *Design Studies* 51 (July): 66–89.
Paavilainen, Heidi, Petra Ahde-Deal, and Ilpo Koskinen. 2017. "Dwelling with Design." *The Design Journal* 20 (1): 13–27.
Papanek, Victor. 1972. *Design for the Real World*. London: Thames and Hudson.
Pateman, Carole. 1970. *Pariticpation and Democratic Theory*. Cambridge University Press.
Patrício, Lia, Raymond P. Fisk, João Falcão e Cunha, and Larry Constantine. 2011. "Multilevel Service Design: From Customer Value Constellation to Service Experience Blueprinting." *Journal of Service Research* 14 (2): 180–200.
Patton, Michael Quinn. 1990. *Qualitative Evaluation and Research Methods*. Newbury Park, CA: Sage Publications.
Pfaffenberger, Bryan. 1992. "Technological Dramas." *Science, Technology, & Human Values* 17 (3): 282–312.

Piller, Frank, and Joel West. 2014. "Firms, Users, and Innovation: An Interactive Model of Couples Open Innovation." *New Frontiers in Open Innovation* 29 (1): 29–49.

Pinto, Nathaly, and Efrén Nango. 2025. "Designing Together in Multi-crisis Times: Effects of Mundane and Strategic Work with Indigenous Communities in the Ecuadorian Amazon." In *Designing Transformative Change: The Potential of Collaboration and Creativity in Crises. Proceedings of the Social Design Network's Conference »On the Verge: Design in Times of Crisis«*, edited by Bori Fehér and Janka Csernák, 197–220.

Pirinen, Antti. 2016. "The Barriers and Enablers of Co-design for Services." *International Journal of Design* 10 (3): 27–42.

Pirinen, Antti, Kaisa Savolainen, Sampsa Hyysalo, and Tuuli Mattelmäki. 2022. "Design Enters the City: Requisites and Points of Friction in Deepening Public Sector Design." *International Journal of Design* 16 (3): 1–19.

Poikolainen Rosén, Anton, Yuxi Chen, Suzanna Törnroth, İdil Gaziulusoy, and Tatu Marttila. 2025. "Four Questions for Systemic More-than-Human Design in Practice." In *More-than-Human Design in Practice*. Taylor & Francis.

Pollock, Neil, and Sampsa Hyysalo. 2014. "The Business of Being a User: The Role of the Reference Actor in Shaping Packaged Enterprise System Acquisition and Development." *MIS Quarterly* 38 (2): 473–96.

Pollock, Neil, and Robin Williams. 2008. *Software and Organisations: The Biography of the Enterprise-Wide System or How Sap Conquered the World*. London: Routledge.

Pollock, Neil, Robin Williams, and Luciana D'Adderio. 2007. "Global Software and Its Provenance: Generification Work in the Production of Organizational Software Packages." *Social Studies of Science* 37: 254–80.

Pollock, Neil, Robin Williams, and Luciana D'Adderio. 2016. "Generification as a Strategy: How Software Producers Configure Products, Manage User Communities and Segment Markets." In *The New Production of Users: Changing Innovation Collectives and Involvement Strategies*, edited by Sampsa Hyysalo, Torben Elgaard Jensen, and Nelly Oudshoorn, 160–92. New York: Routledge.

Prahalad, C., and V. Ramaswamy. 2004a. *The Future of Competition: Co-creating Unique Value with Customers*. Boston: Harward Business School Press.

Prahalad, Coimbatore Krishna, and Venkat Ramaswamy. 2004b. *The Future of Competition: Co-creating Unique Value with Customers*. Boston: Harvard Business Press.

Prandelli, Emanuela, Mohanbir Swahney, and Gianmario Verona. 2008. *Collaborating with Customers to Innovate: Conceiving and Marketing Products in the Networking Age*. Edward Elgar Publishing.

Ramella, Francesco, and Cecilia Manzo. 2018. "Into the Crisis: Fab Labs—A European Story." *The Sociological Review* 66 (2): 341–64.

Rask, M., Andrea, Luciano d', Timo Aarrevaara, Kaisa Matschoss, et al. 2018. *Public Participation, Science and Society*. Taylor & Francis.

Ratto, Matt. 2003. *The Pressure of Opennes: The Hybrid Work of Linux Free/Open Source Kernel Developers*. San Diego: UCSD

Ratto, Matt, and Megan Boler, eds. 2014. *DIY Citizenship: Critical Making and Social Media*. Cambridge, MA: The MIT Press.

Redström, Johan. 2006. "Towards User Design? On the Shift from Object to User as the Subject of Design." *Design Studies* 27: 123–39.

Redström, Johan. 2008. "RE: Definitions of Use." *Design Studies* 29 (4): 410–23.

Renn, Ortwin, Thomas Webler, and Peter Wiedemann, eds. 1995. *Fairness and Competence in Citizen Participation: Evaluating Models for Environmental Discourse*. Vol. 10. Springer Science & Business Media.

Richterich, Annika. 2020. "When Open Source Design Is Vital: Critical Making of DIY Healthcare Equipment During the COVID-19 Pandemic." *Health Sociology Review* 29 (2): 158–67.

Robertson, Suzanne, and James Robertson. 2006. *Mastering the Requirements Process*. Second Edition. Addison Wesley Professional.

Rohracher, Harald. 2005. *User Involvement in Innovation Processes: Strategies and Limitations from a Socio-technical Perspective*.

Roorda, Chris, Niki Frantzeskaki, Derk Loorbach, F. van Steenbergen, and Julia Wittmayer. 2012. *Transition Management in Urban Context: Guidance Manual, Collaborative Evaluation Version*. Rotterdam: Drift, Erasmus University Rotterdam.

Rosch, Eleanor. 1973. "Natural Categories." *Cognitive Psychology* 4 (3): 328–50.

Rosch, Eleanor. 1975. "Cognitive Representations of Semantic Categories." *Journal of Experimental Psychology: General* 104 (3): 192–233.

Ross, Philippe. 2011. "Problematizing the User in User-Centered Production: A New Media Lab Meets Its Audiences." *Social Studies of Science* 41 (2): 251–70.

Rossi, Emilio, Massimo Di Nicolantonio, Fabrizio Ceschin, Giuseppe Mincolelli, Aguinaldo dos Santos, Cindy Kohtala, Edu Fernandes Lima Jacques Filho, Carla Cipolla, and Ezio Manzini. 2021. "Design Contributions for the COVID-19 Global Emergency (Part 2): Methodological Reflections and Future Visions." *Strategic Design Research Journal* 14 (1): 1–18.

Russell, Stewart, and Robin Williams. 2002. "Social Shaping of Technology: Frameworks, Findings and Implications for Policy with Glossary of Social Shaping Concepts." In *Shaping Technology, Guiding Policy: Concepts, Spaces and Tools*, edited by, K. H. Sørensen, and R. Williams, 37–132. Edward Elgar Publishing.

Ryghaug, Marianne, and Tomas Moe Skjølsvold. 2021. *Pilot Society and the Energy Transition: The Co-shaping of Innovation, Participation and Politics*. Springer Nature.

Ryghaug, Marianne, Tomas Moe Skjølsvold, and Sara Heidenreich. 2018. "Creating Energy Citizenship Through Material Participation." *Social Studies of Science* 48 (2): 283–303.

Salovaara, Antti. 2012. "Repurposive Appropriation and Creative Technology Use in Human–Computer Interaction." Academic PhD Dissertation, Helsinki: University of Helsinki.

Sanders, Elizabeth B.-N. 2006. "Scaffolds for Building Everyday Creativity." In *Design for Effective Communications: Creating Contexts for Clarity and Meaning*, edited by Jorge Frascara. New York, NY: Allworth Press.

Sanders, Elizabeth B.-N., and Pieter Jan Stappers. 2008. "Co-creation and the New Landscapes of Design." *CoDesign* 4 (1): 5–18.

Sanders, Elizabeth B.-N., and Pieter Jan Stappers. 2012. *Convivial Toolbox: Generative Research for the Front End of Design*. Amsterdam: BIS.

Sanders, Elizabeth B.-N., and Pieter Jan Stappers. 2014. "Probes, Toolkits and Prototypes: Three Approaches to Making in Codesigning." *CoDesign* 10 (1): 5–14.

Savolainen, Kaisa. 2021. "User-Centred Design Without Involving Users: A Longitudinal Case Study in a Human-Centred-Design–Mature Company." *The Design Journal* 24 (6): 887–905.

Savolainen, Kaisa, and Sampsa Hyysalo. 2020. "Cumulative and Combined: Analyzing Methods Use in a Human-Centered Design Mature Company." *Journal of Usability Studies* 15 (2): 103–23.

Savolainen, Kaisa, and Sampsa Hyysalo. 2021. "User Representations as a Design Resource: Representational Work in Achieving Accountable Design Without Access to Users." *Science & Technology Studies* 34 (2): 25–45.

Sawhney, Mohanbir, Gianmario Verona, and Emanuela Prandelli. 2005. "Collaborating to Create: The Internet as a Platform for Customer Engagement in Product Innovation." *Journal of Interactive Marketing* 19 (4): 4–17.

Schiavone, Francesco, and Salvatore Esposito De Falco. 2016. "Limits to Creative Destruction and Technologies in Practice: The Case of Ham Radio." *Technology Analysis & Strategic Management* 28 (1): 60–75.

Schön, Donald A. 2017. *The Reflective Practitioner: How Professionals Think in Action*. London: Routledge.

Schrage, Michael. 1999. *Serious Play: How the World's Best Companies Simulate to Innovate*. Boston: Harvard Business Press.

Schuler, Douglas, and Aki Namioka. 1993. *Participatory Design: Principles and Practices*. CRC Press.

Schweisfurth, Tim G., and Christina Raasch. 2015. "Embedded Lead Users—The Benefits of Employing Users for Corporate Innovation." *Research Policy* 44 (1): 168–80.

Sengers, Frans, Bruno Turnheim, and Frans Berkhout. 2021. "Beyond Experiments: Embedding Outcomes in Climate Governance." *Environment and Planning C* 39 (6): 1148–71.

Seravalli, Anna. 2014. "While Waiting for the Third Industrial Revolution: Attempts at Commoning Production." In *Making Futures: Marginal Notes on Innovation, Design and Democracy*, edited by Pelle Ehn, Elisabet M. Nilsson, and Richard Topgaard, 99–129. Cambridge, MA: The MIT Press.

Sharrock, Wes, and Bob Anderson. 1994. "The User as a Scenic Feature of the Design Space." *Design Studies* 15 (1): 5–18.

Shove, Elizabeth, Matthew Watson, Martin Hard, and Jack Ingram. 2007. *The Design of Everyday Life*. Oxford: Berg.

Shove, Elizabeth, Matthew Watson, and Mika Pantzar. 2012. *The Dynamics of Social Practice: Everyday Life and How It Changes*, 1–208. London: Sage.

Silverstone, Roger, Eric Hirsch, and David Morley. 1992. "Information and Communication Techonologies and the Moral Economy of the Household." In *Consuming Technologies*, edited by Roger Silverstone and Eric Hirsch, 15–32. London: Routledge.

Silvikko de Villafranca, Marika, Sini Numminen, and Sampsa Hyysalo. 2025. "Characterizing Hybrid Heating in the Households: Diverse Configurational Arrangements Premised on Citizen's Agency and Peer-Support." *Environmental Innovation and Societal Transitions* 55 (June): 100958.

Simon, Herbert Alexander. 1996. *The Sciences of the Artificial*. Cambridge, MA: MIT Press.

Simon, Nina. 2010. *The Participatory Museum*. Santa Cruz: Museum 2.0.

Simonsen, Jesper, and Toni Robertson, eds. 2013. *Routledge International Handbook of Participatory Design*. Vol. 711. New York: Routledge.

Simonsen, Jesper, Helena Karasti, and Morten Hertzum. 2020. "Infrastructuring and Participatory Design: Exploring Infrastructural Inversion as Analytic, Empirical and Generative." *Computer Supported Cooperative Work (CSCW)* 29 (1–2): 115–51.
Sivek, Susan Currie. 2011. "'We Need a Showing of All Hands': Technological Utopianism in *MAKE* Magazine." *Journal of Communication Inquiry* 35 (3): 187–209.
Smith, Adrian. 2012. "Civil Society in Sustainable Energy Transitions." In *Governing the Energy Transition*, 180–202. Routledge.
Smith, Adrian, and Rob Raven. 2012. "What Is Protective Space? Reconsidering Niches in Transitions to Sustainability." *Research Policy*, Special Section on Sustainability Transitions, 41 (6): 1025–36.
Smith, Adrian, Mariano Fressoli, and Hernán Thomas. 2014. "Grassroots Innovation Movements: Challenges and Contributions." *Journal of Cleaner Production*, Special Volume: Sustainable Production, Consumption and Livelihoods: Global and Regional Research Perspectives, 63 (January): 114–24.
Smith, Rachel Charlotte, and Ole Sejer Iversen. 2018. "Participatory Design for Sustainable Social Change." *Design Studies*, Participatory Design, 59 (November): 9–36.
Smith, Rachel Charlotte, Daria Loi, Heike Winschiers-Theophilus, Liesbeth Huybrechts, and Jesper Simonsen. 2025a. *Routledge International Handbook of Contemporary Participatory Design*. Abingdon: Taylor & Francis.
Smith, Rachel Charlotte, Liesbeth Huybrechts, Heike Winschiers-Theophilus, Daria Loi, and Jesper Simonsen. 2025b. "Contemporary Participatory Design: An Introduction." In *Routledge International Handbook of Contemporary Participatory Design*, 1–23. Routledge.
Smith, Thomas S. J. 2017. "Of Makerspaces and Hacklabs: Emergence, Experiment and Ontological Theatre at the Edinburgh Hacklab, Scotland." *Scottish Geographical Journal* 133 (2): 130–54.
Smith, Thomas S. J. 2020. "'Stand Back and Watch Us': Post-capitalist Practices in the Maker Movement." *Environment and Planning A* 52 (3): 593–610.
Sørensen, Knut Holtan, and Robin Williams. 2002. *Shaping Technology, Guiding Policy: Concepts, Spaces and Tools*. Cheltenham, UK: Edward Elgar.
Sovacool, Benjamin K. 2016. "How Long Will It Take? Conceptualizing the Temporal Dynamics of Energy Transitions." *Energy Research & Social Science*, Energy Transitions in Europe: Emerging Challenges, Innovative Approaches, and Possible Solutions, 13 (March): 202–15.
Spekkink, Wouter, Malte Rödl, and Martin Charter. 2022. "Repair Cafés and Precious Plastic as Translocal Networks for the Circular Economy." *Journal of Cleaner Production* 380 (December): 135125.
Staley, David J., Scott Seaman, and Eileen Theodore-Shusta. 2012. "Futuring, Strategic Planning and Shared Awareness: An Ohio University Libraries' Case Study." *The Journal of Academic Librarianship* 38 (1): 1–5.
Star, Susan Leigh. 1995. *Ecologies of Knowledge: Work and Politics in Science and Technology*. Albany: State University of New York Press.
Star, Susan Leigh, and James R. Griesemer. 1989. "Institutional Ecology, 'Translations' and Boundary Objects: Amateurs and Professionals in Berkeley's Museum of Vertebrate Zoology, 1907–39." *Social Studies of Science* 19 (3): 387–420.

Star, Susan Leigh, and Karen Ruhleder. 1996. "Steps Toward an Ecology of Infrastructure: Design and Access for Large Information Spaces." *Information Systems Research* 7 (1): 111–34.

Star, Susan Leigh, and Anselm Strauss. 1999. "Layers of Silence, Arenas of Voice: The Ecology of Visible and Invisible Work." *Computer Supported Cooperative Work (CSCW)* 8 (1): 9–30.

Steen, Marc. 2011. "Tensions in Human-Centred Design." *CoDesign* 7 (1): 45–60.

Steen, Marc, Menno Manschot, and Nicole De Koning. 2011. "Benefits of Co-design in Service Design Projects." *International Journal of Design* 5 (2): 53–60.

Stewart, James, and Sampsa Hyysalo. 2008. "Intermediaries, Users and Social Learning in Technological Innovation." *International Journal of Innovation Management* 12 (3): 295–325.

Strauss, Anselm Leonard. 1978. "A Social World Perspective." *Studies in Symbolic Interaction* 1 (1): 119–28.

Strauss, Anselm Leonard. 2008. *Continual Permutations of Action*. New York: Aldine de Gruyter.

Suchman, Lucy. 1987. *Plans and Situated Actions: The Problem of Human–Machine Communication*. Cambridge: Cambridge University Press.

Suchman, Lucy. 1995. "Making Work Visible." *Communications of the ACM* 38 (9): 56–64.

Szymanski, Margaret H., and Jack Whalen. 2011. *Making Work Visible: Ethnographically Grounded Case Studies of Work Practice*. Cambridge: Cambridge University Press.

Taffe, Simone. 2015. "The Hybrid Designer/End-User: Revealing Paradoxes in Co-design." *Design Studies* 40 (September): 39–59.

Thrift, Nigel. 2006. "Re-inventing Invention: New Tendencies in Capitalist Commodification." *Economy and Society* 35 (2): 279–306.

Toombs, Austin, Shaowen Bardzell, and Jeffrey Bardzell. 2014. "Becoming Makers: Hackerspace Member Habits, Values, and Identities." *Journal of Peer Production* 5 (2014): 1–8.

Tooze, James, Sharon Baurley, Robert Phillips, Paul Smith, Edwin Foote, and Sarah Silve. 2014. "Open Design: Contributions, Solutions, Processes and Projects." *The Design Journal* 17 (4): 538–59.

Törpel, Bettina, Alex Voss, Mark Hartswood, and Rob Procter. 2009. "Participatory Design: Issues and Approaches in Dynamic Constellations of Use, Design, and Research." In *Configuring User-Designer Relations*, 13–29. London: Springer.

Torvinen, Vesa. 1999. *Construction and Evaluation of the Labour Game Method*. Turku Centre for Computer Science.

Toupin, Sophie. 2014. "Feminist Hackerspaces: The Synthesis of Feminist and Hacker Cultures." *Journal of Peer Production* 5: 1–11.

Turnhout, Koen van, Arthur Bennis, Sabine Craenmehr, Robert Holwerda, Marjolein Jacobs, Ralph Niels, Lambert Zaad, Stijn Hoppenbrouwers, Dick Lenior, and René Bakker. 2014. "Design Patterns for Mixed-Method Research in HCI." In *Proceedings of the 8th Nordic Conference on Human–Computer Interaction: Fun, Fast, Foundational*, 361–70. NordiCHI '14. New York, NY, USA: Association for Computing Machinery.

Turnhout, Esther, Tamara Metze, Carina Wyborn, Nicole Klenk, and Elena Louder. 2020. "The Politics of Co-production: Participation, Power, and Transformation." *Current Opinion in Environmental Sustainability* 42: 15–21.

Urban, Glen L., and Eric von Hippel. 1988. "Lead User Analyses for the Development of New Industrial Products." *Management Science* 34 (5): 569–82.
Usenyuk, Svetlana, Sampsa Hyysalo, and Jack Whalen. 2016. "Proximal Design: Users as Designers of Mobility in the Russian North." *Technology and Culture* 57 (4): 866–908.
Usenyuk-Kravchuk, Svetlana, Sampsa Hyysalo, and Alexandra Raeva. 2022. "Local Adequacy as a Design Strategy in Place-Based Making." *CoDesign* 18 (1): 115–34.
Vaajakallio, Kirsikka. 2012. *Design Games as a Tool, a Mindset and a Structure.* Helsinki: Aalto University.
Valtonen, Anna. 2007. "Redefining Industrial Design: Changes in the Design Practice in Finland." PhD diss., University of Art and Design Helsinki.
Valve, Helena, Jani Lukkarinen, Antti Belinskij, Pilvi Kara, Liisa Kolehmainen, Aleksis Klap, Riitta Leskinen, Satu Lähteenoja, Tatu Marttila, and Maiju Oikarinen. 2019. "Lisäarvoa Kalasta Ja Maatalouden Sivuvirroista Varsinais-Suomessa: Sinisen Biotalouden Murrosareenan Tulokset" (Added Value from Fish and Agricultural Side-Streams in South-West of Finland: Results from BlueAdapt Transition Arena).
Valve, Helena, Jani Lukkarinen, Ville Matikka, Harri Auvinen, Satu Lähteenoja, and Tatu Marttila. 2019. "Kohti Kestävän Sinisen Biotalouden Murrosta Pohjois-Savossa: Murrospolut Ja Tarvittavat Toimenpiteet" (Towards a Sustainable Transition in Blue Bioeconomy in North Savo: Transition Pathways and Necessary Actions).
Valve, Helena, David Lazarevic, Sampsa Hyysalo, Jani Lukkarinen, and Tatu Marttila. 2023. "The Interrupting Capacities of Knowledge Co-production Experiments: A Sociology of Testing Approach." *Environmental Science & Policy* 147 (September): 255–64.
Varho, Vilja, Sanne Bor, Kirsi Kataja, Jussi Lahtinen, Taina Lahtinen, Henna Sundqvist, Titta Tapiola, and Maria Åkerman. 2023. *Askelet kestävään ruokapakkaamiseen: Kestävän ruokapakkaamisen murrosareena.* Luonnonvarakeskus. (Steps to Sustainable Food Packaging: Transition Arena of Sustainable Foodpackaging.)
Verhaegh, Stefan, Ellen van Oost, and Nelly Oudshoorn. 2016. "Innovation in Civil Society: The Socio-Material Dynamics of a Community Innovation." In *The New Production of Users: Changing Innovation Collectives and Involvement Strategies*, 193–218. Routledge.
Vermeeren, Arnold P. O. S., Effie Lai-Chong Law, Virpi Roto, Marianna Obrist, Jettie Hoonhout, and Kaisa Väänänen-Vainio-Mattila. 2010. "User Experience Evaluation Methods: Current State and Development Needs." In *Proceedings of the 6th Nordic Conference on Human–Computer Interaction: Extending Boundaries*, 521–30. NordiCHI '10. New York, NY, USA: Association for Computing Machinery.
Vertiz Marquez, Brenda Victoria. 2024. "Designing for Social Reproduction: Towards the Sustainment of Public Space Interventions in Mexico City's Colonias Populares." In *PDC '24: Proceedings of the Participatory Design Conference 2024: Full Papers*, edited by V. D'Andrea, R. A. de Paula, A. A. Geppert, M. Brereton, C. Del Gaudio, M. Y. Jensen, H. Winschiers-Theophilus, and T. Zaman, vol. 1, 14–27. ACM.
Vezzoli, Carlo, Cindy Kohtala, Amrit Srinivasan, J. C. Diehl, Sompit Moi Fusakul, Liu Xin, and Deepta Sateesh. 2017. "Product-Service System Design for Sustainability." In *Product-Service System Design for Sustainability.* Routledge.

Vezzoli, Carlo, Brenda Garcia Parra, and Cindy Kohtala, eds. 2021. *Designing Sustainability for All: The Design of Sustainable Product-Service Systems Applied to Distributed Economies*. Springer Nature.

Victor, Bart, and Andrew C. Boynton. 1998. *Invented Here: Maximizing Your Organization's Internal Growth and Profitability*. A Practical Guide to Transforming Work. Boston, MA: Harward Business School Press.

Visser, Froukje Sleeswijk, Pieter Jan Stappers, Remko van der Lugt, and Elizabeth B.-N. Sanders. 2005. "Contextmapping: Experiences from Practice." *CoDesign* 1 (2): 119–49.

von Busch, Otto, and Karl Palmås. 2023. *The Corruption of Co-design: Political and Social Conflicts in Participatory Design Thinking*. New York: Routledge.

von Hippel, Eric. 1986. "Lead Users: A Source of Novel Product Concepts." *Management Science* 32 (7): 791–805.

von Hippel, Eric. 1994. "'Sticky Information' and the Locus of Problem Solving: Implications for Innovation." *Management Science* 40 (4): 429–39.

von Hippel, Eric. 2001. "Innovation by User Communities." In *Communities of Practice*. Routledge.

von Hippel, Eric. 2006. *Democratizing Innovation*. Cambridge: The MIT Press.

von Hippel, Eric. 2016. *Free Innovation*. Cambridge: The MIT Press.

von Hippel, Eric, and Ralph Katz. 2002. "Shifting Innovation to Users Via Toolkits." *Management Science* 48 (7): 821–33.

von Hippel, Eric, and Georg Von Krogh. 2016. "Crossroads—Identifying Viable 'Need–Solution Pairs': Problem Solving Without Problem Formulation." *Organization Science* 27 (1): 207–21.

Voss, Alex, Mark Hartswood, Rob Procter, Mark Rouncefield, Roger Slack, and Monika Büscher. 2009. *Configuring User-Designer Relations*. Computer Supported Cooperative Work. London: Springer.

Voss, Alex, Rob Procter, Roger Slack, Mark Hartswood, and Mark Rouncefield. 2009. "Design as and for Collaboration: Making Sense of and Supporting Practical Action." In *Configuring User-Designer Relations*, 31–58. London: Springer.

Voß, Jan-Peter, and Nina Amelung. 2016. "Innovating Public Participation Methods: Technoscientization and Reflexive Engagement." *Social Studies of Science* 46 (5): 749–72.

Voß, Jan-Peter, Adrian Smith, and John Grin. 2009. "Designing Long-Term Policy: Rethinking Transition Management." *Policy Sciences* 42 (4): 275–302.

Vygotsky, Lev. 1979. *Thought and Language*. Moscow: Progress.

Wardlaw, Jessica C. 2016. "Introducing User-Centred Design: A Longitudinal Study of a Healthcare Informatics Organisation." Doctoral, UCL (University College London).

Weber, K. Matthias, and Harald Rohracher. 2012. "Legitimizing Research, Technology and Innovation Policies for Transformative Change: Combining Insights from Innovation Systems and Multi-Level Perspective in a Comprehensive 'Failures' Framework." *Research Policy*, Special Section on Sustainability Transitions, 41 (6): 1037–47.

Wenger, Etienne. 1998. *Communities of Practice: Learning, Meaning, and Identity*. Cambridge: Cambridge University Press.

Whalen, Jack, and Daniel G. Bobrow. 2011. "Communal Knowledge Sharing: The Eureka Story." In *Making Work Visible: Ethnographically Grounded Case Studies of Work Practice*, edited by Margaret H. Szymanski, and Jack Whalen, 257–84. Cambridge: Cambridge University Press.

Wigboldus, Seerp, Laurens Klerkx, Cees Leeuwis, Marc Schut, Sander Muilerman, and Henk Jochemsen. 2016. "Systemic Perspectives on Scaling Agricultural Innovations. A Review." *Agronomy for Sustainable Development* 36 (3): 1–20.

Wilkie, Alex. 2010. "User Assemblages in Design: An Ethnographic Study." Doctoral, Goldsmiths, University of London.

Williams, Robin, and David Edge. 1996. "The Social Shaping of Technology." *Research Policy* 25 (6): 865–99.

Williams, Robin, James Stewart, and Roger Slack. 2005. *Social Learning in Technological Innovation: Experimenting with Information and Communication Technologies*. UK: Edward Elgar Publishing.

Woodhouse, Edward, and Jason W. Patton. 2004. "Introduction: Design by Society: Science and Technology Studies and the Social Shaping of Design." *Design Issues* 20 (3): 1–12.

Woolgar, Steve. 1991. "Configuring the User: The Case of Usability Trials." In *A Sociology of Monsters: Essays on Power, Technology, and Domination*, edited by John Law, 58–99. London: Routledge.

Woolgar, Steve, and Javier Lezaun. 2013. "The Wrong Bin Bag: A Turn to Ontology in Science and Technology Studies?" *Social Studies of Science* 43 (3): 321–40.

Woolrych, Alan, Kasper Hornbæk, Erik Frøkjær, and Gilbert Cockton. 2011. "Ingredients and Meals Rather than Recipes: A Proposal for Research that Does Not Treat Usability Evaluation Methods as Indivisible Wholes." *International Journal of Human–Computer Interaction* 27: 940–70.

Wu, Yiying. 2017. "Bicycles and Plants: Designing for Conviviality and Meaningful Social Relations Through Collaborative Services." Doctoral Dissertation, Helsinki, Finland: Aalto University School of Arts, Design and Architecture, Department of Design.

Zimmerman, Eric. 2003. "Play as Research: The Iterative Design Process." *Design Research: Methods and Perspectives* 2003: 176–84.

Author Index

S. Hyysalo, *Design Participation*,
https://doi.org/10.1007/978-3-031-95437-5

U

V

W

Index

S. Hyysalo, *Design Participation*,
https://doi.org/10.1007/978-3-031-95437-5

The manufacturer's authorised representative in the EU is Springer
Nature Customer Service Centre GmbH, Europaplatz 3, 69115 Heidelberg,
Germany. If you have any concerns regarding our products, please
contact ProductSafety@springernature.com

Printed and bound by CPI Group (UK) Ltd, Croydon, CR0 4YY
30/09/2026
02225507-0012